BLOCKCH

TECHNOLOGY AND ITS

APPLICATIONS

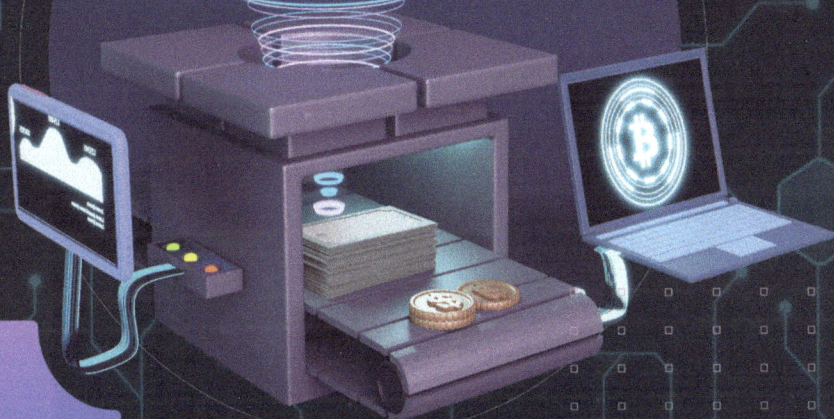

Matthew N. O. Sadiku

BLOCKCHAIN TECHNOLOGY

AND ITS

APPLICATIONS

Matthew N. O. Sadiku, Ph.D., P.E.

Regents Professor Emeritus and IEEE Life Fellow
Prairie View A&M University
Prairie View, TX 77446

Date published: 2023

Matthew N. O. Sadiku
sadiku@ieee.org
www.matthew-sadiku.com

Paperback: 979-8-9890916-7-6
eBook: 979-8-9890916-8-3

BookFilmsMedia
2780 South Jones Blvd Suite 200- 4007
Las Vegas, NV 89146 United States
+1 725-238-6534

DEDICATED TO MY WIFE,

JANET O. SADIKU

BRIEF TABLE OF CONTENTS

PREFACE

Advances in technology always have an impact on our society. Emerging technologies, such as the Internet of things (IoT), artificial intelligence, and Blockchain, present transformative opportunities. The 21st century is all about revolutionizing technologies. One of the leading technologies that has changed many aspects is Blockchain. Blockchain came as a solution to the longstanding user's trust problem. It is a technology that builds a trustworthy service in an untrustworthy environment. It refers to a highly secure and decentralized ledger system on which information can be stored but cannot be altered. It has evolved beyond cryptocurrencies to general purpose and can be used across an array of applications. It has proven to be another revolutionary technology that will impact many industries and transactions.

Blockchain (also known as "distributed ledger technology") is a peer-to-peer network that sits on top of the Internet. Blockchain technology is an innovation which is regarded as the center of Industry 4.0 revolution and it has become part of our lives. It is a system that stores data in a special way. Blockchain technology has some interesting properties, such as its decentralized nature, immutability, decentralization, transparency, and permissionless, that may be used to address pressing issues in many sectors. Although this technology finds its first application in the financial sector, it has become possible to use it in all sectors which can be integrated with technology today.

Before Blockchain technology, people turned to gold or real estate when inflation hit its peak. Today, governments all over the globe have started opening up to Blockchain and crypto. By using blockchain, governments can reduce administrative costs, increase transparency, and improve service delivery. Blockchain is revolutionizing the digital world by bringing a new perspective to security, efficiency, and stability of systems and data. It is network of computers that is decentralized. Blockchain keeps track of distributed data and provides encrypted transaction tracking. It has attracted attention with its unique characteristics, such as irrevocability and security. It will be a part of our everyday life.

This book explores Blockchain and its various applications. It is organized into fourteen chapters that summarize Blockchain technology and its applications.

Chapter 1: Introduction:

This chapter introduces Blockchain and serves an introduction to this book. It explains how its unique characteristics could help fundamentally change the building blocks of commerce and society. Bitcoin is a cryptographic electronic payment system that purports to be the world's first cryptocurrency. The Blockchain could bring everything that is good about Bitcoin and translate it into decentralized applications. Blockchain, a type of distributed digital ledger technology, is a relatively new and exciting way of recording transactions in the digital age.

Chapter 2: Blockchain in Business:

This chapter is an introduction to the use of Blockchain technology in business. The business world is constantly evolving, and technological advancements play a pivotal role in shaping its landscape. Traditional business models are no longer sufficient to tackle global competition due their inherent limitations. Blockchain is one of the major technologies driving business transformation. While Blockchain holds immense potential for revolutionizing various aspects of business operations, it is important for businesses to carefully assess their needs, understand the technology, and consider the challenges before implementing Blockchain solutions.

Chapter 3: Blockchain in Finance:

The chapter provides a primer on the use of Blockchain technology in finance. Finance industry is one of the largest booming sectors. It was the first sector to which Blockchain (BC) technology was applied. This sector witnesses millions of transactions worth trillions of dollars daily. Blockchain is basically a ledger of recorded financial transactions. It provides a potentially attractive alternative way to organize modern finance. It facilitates safe, easy transactions, and builds trust between trading partners. When trust in the central hubs of finance is being increasingly questioned, decentralized systems like the Blockchain that reduce the need for such trust become attractive.

Chapter 4: Blockchain in Healthcare:

In this chapter, we consider the use of Blockchain healthcare. The healthcare industry is one of the world's largest industries and is resistant to change and innovative practices. As a catalyst for change, the Blockchain technology is going to change healthcare in major ways.

The main motivation for using Blockchain in healthcare is to solve the data integrality, data interoperability, and privacy issues in current health systems. Blockchain eliminates the need of a middleman who plays the role of verifying transaction in the healthcare industry.

Chapter 5: Blockchain in Agriculture:

This chapter examines the applications of Blockchain technology in agriculture. Agriculture is the production of food, fiber, and data. It has played a crucial role in the promotion of life and wellbeing around the world. Blockchain has different uses in the agricultural industry such as providing solutions to food safety, food waste, food fraud, supply chain visibility, and management. Several farmers and agribusinesses have started introducing Blockchain technology in agriculture.

Chapter 6: Blockchain in Supply Chain:

In this chapter, we introduce the integration of Blockchain in supply chain. Supply chain is a set of sequential stages in the manufacturing, transportation, storing, or distribution of a product. It plays a critical role in the global economy. Today's supply chains are global networks that generally include manufacturers, suppliers, logistics companies, and retailers that work together to deliver products to consumers. They operate without Blockchain technology. Several prominent, forward-thinking companies are testing Blockchain solutions and investigating Blockchain uses for their supply chains. Blockchain supports supply chain management by resolving some of the existing concerns, like provenance, transparency, performance improvement, quality assurance and control, and achieving sustainability.

Chapter 7: Blockchain in Government:

This chapter provides an overview on blockchain in government. Most government services around the world run on inefficient systems loaded with heavy bureaucracy. They lead to non-transparent systems and a loss of public confidence in government services. Blockchain provides governments with a fast, secure, efficient, speedy, trustworthy, and transparent way to deliver government services and communicate with their citizens. It can help build citizens' trust, prevent data breaches, reduce corruption, and cut government spending. As a result, the governments of many countries have expressed interest in the technology.

Chapter 8: Blockchain in Internet of Things:

This chapter provides a short review of the union of the Internet of things (IoT) and Blockchain. The IoT connects people, places, and products, offering new opportunities to generate value in products and business processes. IoT is a network of connected devices and people, collecting and sharing data. IoT holds the promise of connecting almost any device to the Internet and making the devices smarter and more accessible. Integrating Blockchain with the Internet of things can deliver numerous benefits, such as improving transparency, traceability, reliability, and automation. The merging of Blockchain and the IoT is also expected to address the issue security and privacy.

Chapter 9: Blockchain in Smart Cities:

This chapter explores the potential of Blockchain as a technology for enabling smart cities. A smart city is a high-tech urban area that connects people, information, and technologies in order to increase life quality. It integrates critical infrastructures such as bridges, tunnels, roads, subways, airports, seaports, and buildings in a secure manner so as to manage the city's assets and improve the efficiency of services such as energy, water, and transportation. Cities around the world are already working becoming "smart cities." They are using Blockchain technology to enhance their operations and services. Applying Blockchain technology to smart cities produces several benefits such as trust-free, transparency, pseudonymity, democracy, automation, decentralization, and security.

Chapter 10: Blockchain in Cybersecurity:

This chapter provides a primer on Blockchain security. Cybersecurity is the practice of protecting systems and networks from digital attacks. Blockchain is the latest cybersecurity technology that is gaining popularity and recognition. Blockchain provides security, anonymity, and data integrity without the need of a third party. Cybersecurity is built into Blockchain technology because of its inherent nature of being a decentralized system based on principles of security, privacy, and trust. Blockchain technology is emerging as the ultimate weapon in the fight against cybercrime.

Chapter 11: Blockchain in Social Media:

In this chapter, we consider the use of Blockchain technology in social media. Social media, such as Facebook, Twitter, Instagram, Youtube, Whatsapp, Snapchat, etc. have changed our way of communication. They have become the center of the modern Internet.

Modern social networks wield enormous power in society as it is evident by the fact that people are spending significant portions of their lives online. We turn to social media to form groups, foster relationships, and keep in touch with long-distance friends. Blockchain technology offers a promising solution for a new social media architecture. The decentralized nature of Blockchain allows for the creation of platforms where users control their data and have a stake in the network. This eliminates the need for a central authority or intermediary, promoting transparency and fairness.

Chapter 12: Blockchain in Artificial Intelligence:

The chapter considers the combination of AI and Blockchain and its applications. Artificial intelligence (AI) is a technology that can perform complex tasks that require human intelligence, and it holds the potential of exceeding human capabilities. Blockchain and artificial intelligence (AI) are important disruptive, emerging technologies. When both of them are combined, they can help you build an immutable, safe, and decentralized system. The amalgamation of AI and Blockchain holds great potential to create new business models and used for process improvement and value creation such as asset management, customer service, dispute resolution, fraud prevention, production evaluation, and supply chain monitoring.

Chapter 13: Blockchain Around the World:

This chapter summarizes the current Blockchain adoption and regulation in many countries. Global interest in Blockchain technologies and their possible impact have permeated the public consciousness. Due to several factors existing in each country around the world, there are countries that are more ahead of the curve in terms of Blockchain adoption. A practical usage of cryptocurrency around the globe is what differentiates it from every other major currency whether it is euro, dollar, or any other. Numerous country-based financial organizations are putting resources into Blockchain technology to make a more productive organization to deal with economic exchanges.

Chapter 14: Future of Blockchain:

This last chapter attempts to predict the future of Blockchain technology. It explores the future of Blockchain by first considering novel technologies with the potential to facilitate future development of Blockchain, and then delving into its implications in different applications in the future of Blockchain. Such technologies include

Internet of things and artificial intelligence. Although it is difficult to predict the future technology landscape and exactly how Blockchain fits in, but this chapter has identified a few areas of potential impact that will dictate the future of Blockchain.

This is a comprehensive introductory text on the issues, ideas, theories, and problems on Blockchain. It provides an overview on each of its applications so that beginners can understand Blockchain, its increasing importance, and its applications. It is a must-read book for anyone who wants to learn about Blockchain, which is undoubtedly one of the most important technologies that has emerged in the last decade.

I am grateful for the support of Dr. Annamalia Annamalai, the department head of the Department of Electrical and Computer Engineering, and Dr. Pamela Obiomon, the dean of the College of Engineering at Prairie View A&M University, Prairie View, Texas. Special thanks are due to my wife Janet Sadiku for helping with the quotations at the beginning of each chapter.

— M. N. O. Sadiku

ABOUT THE AUTHOR

Matthew N. O. Sadiku:

He received his B. Sc. degree in 1978 from Ahmadu Bello University, Zaria, Nigeria and his M.Sc. and Ph.D. degrees from Tennessee Technological University, Cookeville, TN in 1982 and 1984 respectively. In total, he received seven college degrees. From 1984 to 1988, he was an assistant professor at Florida Atlantic University, Boca Raton, FL, where he did graduate work in computer science. From 1988 to 2000, he was at Temple University, Philadelphia, PA, where he became a full professor. From 2000 to 2002, he was with Lucent/Avaya, Holmdel, NJ as a system engineer and with Boeing Satellite Systems, Los Angeles, CA as a senior scientist. He is presently a Regents professor emeritus of electrical and computer engineering at Prairie View A&M University, Prairie View, TX.

He is the author of over 1,200 professional papers and over 120 books including "Elements of Electromagnetics" (Oxford University Press, 7th ed., 2018), "Fundamentals of Electric Circuits" (McGraw-Hill, 7th ed.,2020, with C. Alexander), "Computational Electromagnetics with MATLAB" (CRC Press, 4th ed., 2019), "Principles of Modern Communication Systems" (Cambridge University Press, 2017, with S. O. Agbo), and "Emerging Internet-based Technologies" (CRC Press, 2019). In addition to the engineering books, he has written Christian books including "Secrets of Successful Marriages," "How to Discover God's Will for Your Life," and commentaries on all the books of the New Testament Bible. Some of his books have been translated into French, Korean, Chinese (and Chinese Long Form in Taiwan), Italian, Portuguese, and Spanish.

He was the recipient of the 2000 McGraw-Hill/Jacob Millman Award for outstanding contributions in the field of electrical engineering. He was also the recipient of Regents Professor award for 2012-2013 by the Texas A&M University System. He is a registered professional engineer and a life fellow of the Institute of Electrical and Electronics Engineers (IEEE) "for contributions to computational electromagnetics and engineering education." He was the IEEE Region 2 Student Activities Committee Chairman. He was an associate editor for IEEE Transactions on Education. He is also a member of Association for Computing Machinery (ACM) and American Society of Engineering Education (ASEE). His current research interests are in the areas of computational electromagnetic, computer science/networks, engineering education, and marriage counseling. His works can be found in his autobiography, "My Life and Work" (Trafford Publishing, 2017) or his website: www.matthew-sadiku.com. He currently resides with his wife Janet in West Palm Beach, Florida. He can be reached via email at sadiku@ieee.org

DETAILED TABLE OF CONTENTS

CHAPTER 1

INTRODUCTION

"When decentralized blockchain protocols start displacing the centralized web services that dominate the current Internet, we'll start to see real internet-based sovereignty. The future Internet will be decentralized."

— Olaf Carlson-Wee

1.1 INTRODUCTION

Contracts, transactions, and their records are critical, defining structures in our economic, legal, and political systems, but they have not being able to keep up with the world's digital transformation. Blockchain (BC) promises to solve this problem. Blockchain (also known as "distributed ledger technology") is a peer-to-peer network that sits on top of the Internet. It was introduced in 2008 as part of a proposal for Bitcoin. Bitcoin is the first application of BC technology. But what is Bitcoin?

Bitcoin is a cryptographic electronic payment system that purports to be the world's first cryptocurrency. It has become the most talked about cryptocurrency. The software is completely open source so that any developer can download it, modify it, and create his own version of the software. This unique feature has led to an explosion of alternative bitcoin implementations, popularly known as altcoins. Some of the popular implementations include IxCoin, Namecoin, Litecoin, Ripple, Dogecoin, and Bitcoin. Some of the key benefits of Bitcoin include security, transparency, lower transaction costs, anonymity, and resilience. Although Bitcoin is a revolutionary idea, its implementation suffers some problems such as instability, deflation, lack of replicability, computational inefficiency, and lack of regulation or enforcement [1].

The Blockchain could bring everything that is good about Bitcoin and translate it into decentralized applications. Blockchain refers to new applications of a distributed database technology that builds on a tamper-proof records of time-stamped transactions.

By decentralizing it, Blockchain makes data transparent to everyone involved and this eliminates the risks that come with data being held centrally. A Blockchain facilitates secure online transactions [2].

This chapter introduces Blockchain and serves an introduction to this book. It explains how its unique characteristics could help fundamentally change the building blocks of commerce and society. It begins with discussing what Blockchain is all about and how it works. It provides some real world applications of BC. It highlights the benefits and challenges of BC. The last section concludes with comments.

1.2 WHAT IS BLOCKCHAIN?

Blockchain, a type of distributed digital ledger technology (DLT), is a relatively new and exciting way of recording transactions in the digital age. It is a decentralized and distributed digital ledger technology that securely records and verifies transactions across multiple computers or nodes in a network. Basically, it is a chain of blocks in which each block contains a list of transactions. The blockchain technology was created as the foundational basis for Bitcoin – a digital currency in which secure peer-to-peer transactions occur over the Internet. It is expected that the spending on blockchain solutions worldwide would grow from 4.5 billion USD (2020) to an estimated value of 19 billion USD by 2024 [3].

Originally developed as the accounting method for the virtual currency Bitcoin, Blockchains are appearing in a variety of commercial applications today. Blockchain technology is a type of distributed digital ledger that uses encryption to make entries permanent and tamper-proof and can be programmed to record financial transactions. It is used for secure transfer of money, assets, and information via a computer network such as the Internet without requiring a third-party intermediary. It is now being adopted across financial and non-financial sectors. As a catalyst for change, the Blockchain technology is going to change the business world and financial matters in major ways.

The first Blockchain was conceived in 2008 by an anonymous person or group known as Satoshi Nakamoto, who published a white paper introducing the concept of a peer-to-peer electronic cash system he called Bitcoin [4,5]. Bitcoin and Ethereum are the first two mainstream Blockchains. Other modern Blockchains include Namecoin, Peercoin, Ether, and Litecoin. Figure 1.1 shows different components of Blockchain [6].

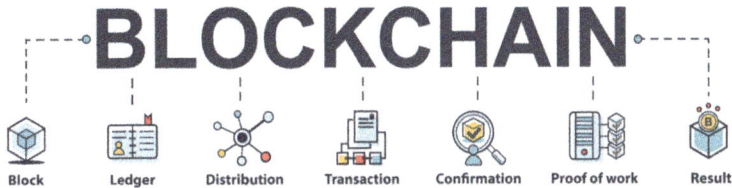

Figure 1.1 Different components of Blockchain [6].

Blockchain combines existing technologies such as distributed digital ledgers, encryption, immutable records management, asset tokenization and decentralized governance to capture and record information that participants in a network need to interact and transact. As illustrated in Figure 1.2, a complete blockchain incorporates all the following five elements [7]:

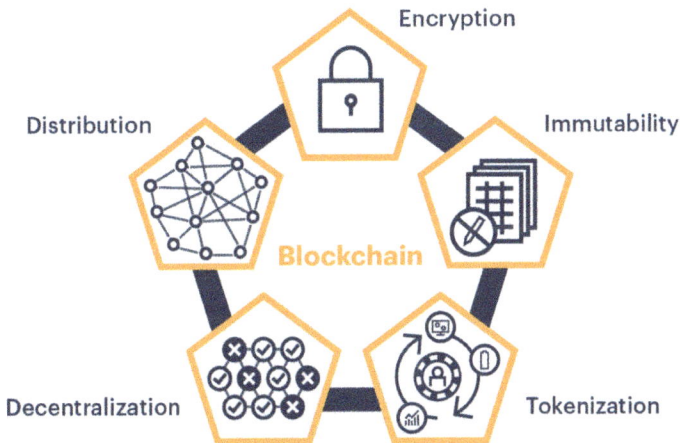

Figure 1.2   Five key elements of Blockchain [7].

- *Distribution*: Digital assets are distributed, not copied or transferred. A protocol establishes a set of rules in the form of distributed mathematical computations that ensures the integrity of the data exchanged among a large number of computing devises without going though a trusted third party. A centralized architecture presents several issues including a single point of failure and problems of scalability.

- *Encryption*: BC uses technologies such as public and

private keys to record data securely and semi-anonymously. Completed transactions are cryptographically signed, time-stamped, and sequentially added to the ledger.

• *Immutability*: The Blockchain was designed so these transactions are immutable, i.e. they cannot be deleted. No entity can modify the transaction records. Thus, Blockchains are secure and meddle-free by design. Data can be distributed, but not copied.

• *Tokenization*: Value is exchanged in the form of tokens, which can represent a wide variety of asset types, including monetary assets, units of data or user identities.

• *Decentralization*: No single entity controls a majority of the nodes or dictates the rules. A consensus mechanism verifies and approves transactions, eliminating the need for a central intermediary to govern the network.

1.3 HOW BLOCKCHAIN WORKS

Blockchain is a distributed ledger technology that evolved from the Internet of information and represents a second phase of the Internet. It is somewhat similar to spreadsheets or databases because it is a database where information is entered and stored. BC is a decentralized form of record-keeping. The key difference between a traditional database (or spreadsheet) and a blockchain is how the data is structured and accessed.

The term "Blockchain" refers to the way BC stores transaction data – in "blocks" that are linked together to form a "chain." The chain grows as the number of transactions increases. A block is created whenever a transaction is made. Each transaction, referred to as a "block," is secured through cryptography, timestamped, and validated by every authorized member of the database using consensus algorithms. Every transaction is attached to the previous transaction in sequential order, creating a chain of transactions (or blocks), as shown in Figure 1.3 [8].

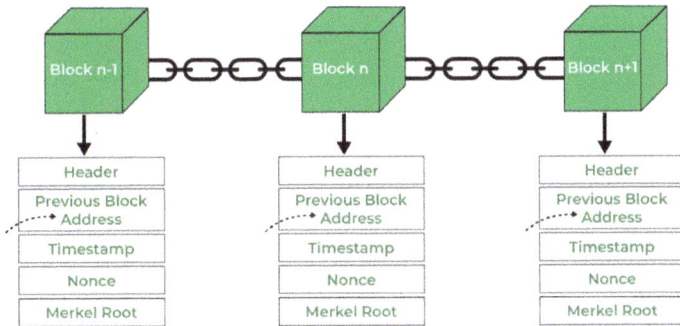

Figure 1.3 A chain of transactions (or blocks) [8].

In other words, a block is the "current" part of a Blockchain, which records some or all of the recent transactions. Each BC blocks has a unique 32-bit whole number called a nonce, which is connected to a 256-bit hash number attached to it. The block is broadcasted to all nodes for validation. Once completed, a block goes into the Blockchain as a permanent database. Each time a block gets completed, a new one is generated. Each data item in a BC has a timestamp. A BC is an ordered chain of blocks. All data of a transaction are traceable based on the chain structure of BC. Figure 1.4 displays how BC works [9].

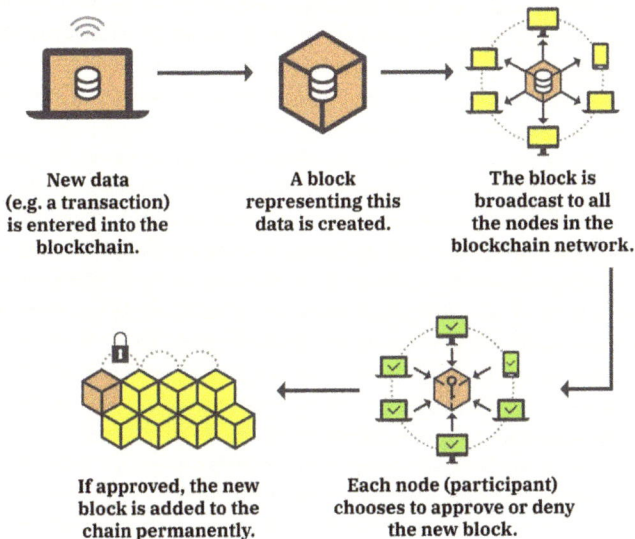

Figure 1.4 How Blockchain works [9].

The BC technology currently has the following features [10,11]:

1. *Peer-to-Peer (P2P) Network*: The first requirement of BC is a network, an infrastructure shared by multiple parties. This can be a LAN at a small scale or the Internet at a large scale. Communication occurs directly between peers instead of through a central node. All nodes participating in a BC are connected in a decentralized P2P network. Transactions are broadcast to the P2P network. Due to some limitations of P2P networks, some vendors have provided cloud-based BCs.

2. *Cascaded Encryption*: A BC uses encryption to protect transaction data. Blocks are encrypted in a cascaded manner, i.e. the encryption result of the previous block is used in encrypting the current block. The BC is secured by public key cryptography, with each peer generating its own public-private key pairs.

3. *Distributed Database*: A BC is digitally distributed across a number of computers. Each party on a BC has access to the entire database and no single party controls the data or the information. Since BC is decentralized, there is no need for central authorizes such as banks.

4. *Transparency with Pseudonymity*: Each node or participant on a blockchain has a unique 30-plus-character alphanumeric address that identifies it. Users can choose to remain anonymous or provide proof of their identity to others.

5. *Irreversibility of Records*: Once a transaction is entered in the database and the accounts are updated, the records cannot be altered. Records on the database is permanent, chronologically ordered, and available to all others on the network.

As displayed in Figure 1.5, there are different types [12].

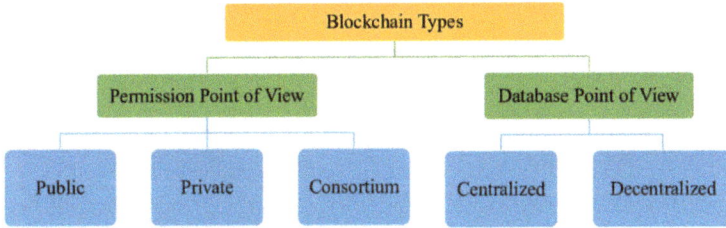

Figure 1.5 *Different types of Blockchain [12].*

The most popular types are public and private. Public Blockchains are cryptocurrencies such as Bitcoin, enabling peer-to-peer transactions. Private Blockchains use Blockchain-based platforms such as Ethereum or Blockchain-as-a-service (BaaS) platforms running on private cloud infrastructure. They limit access to the predefined list of known individuals. A private BC is an intranet, while a public BC is the Internet. Companies will be disrupted the most by public Blockchains.

Just like any technology, BC went through stages and evolved as it progressed and matured. We started with Blockchain 1.0 and now we are at Blockchain 4.0. The stages are shown in Figure 1.6 [13].

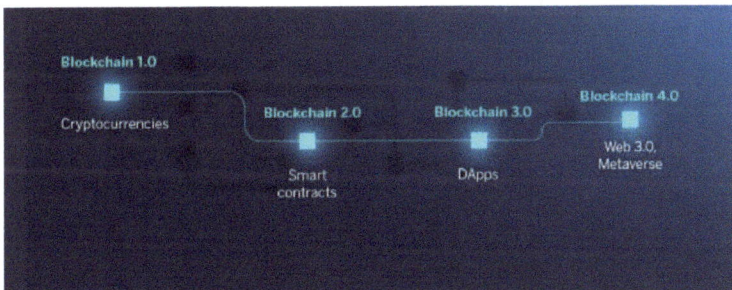

Figure 1.6 *Stages of Blockchain [13].*

BCs may be permissioned or permissionless. In a permissioned BC, each participant has a unique identity. Permissionless BCs allow anyone to join, participate or leave the protocol execution without seeking permission from a centralized or distributed authority.

1.4 APPLICATIONS

Most applications of BC involve cross-organizational business processes, exploiting the neutral ground provided by a BC. Blockchain has been applied by government and non-government organizations. Some of its cutting-edge applications are provided next.

- *Cryptocurrency*: This is also known as digital currency. It is the most well-known of all blockchain applications. Each cryptocurrency has its own, unique blockchain where transactions are combined into blocks and then linked together. For example, the Bitcoin blockchain and Ethereum blockchain do not interact. The cryptographic nature of blockchain networks minimizes the risk of your financial information or identity being compromised, allowing for anonymous and more secure transactions. Presently, over 1,600 digital currencies using blockchain are in circulation [9]. Cryptocurrency can take a company into previously untapped developing regions. Of course, it also can simplify commerce right here at home. Figure 1.7 displays coins of various cryptocurrencies [14].

Figure 1.7   Coins of various cryptocurrencies [14].

- *Smart contracts*: Smart contracts are often regarded as the killer application of Blockchain. BC can be used to create smart contracts which can be executed or enforced without human interaction. Smart contracts are self-executing contracts with the terms of the agreement between buyer and seller being directly written into lines of code. Such contracts permit trusted, transparent, and irreversible transactions and agreements to be carried out among disparate, anonymous parties without the middleman. They turn BC into a middleman to execute all manner of complex business deals, legal agreements, and automated exchanges of data. An example of smart contract utilization is in the music industry, musicians can share free-trade music and ensure that the profits go back to the artists. If contracts are automated through Blockchain, what will happen to traditional structures, processes, and intermediaries like lawyers and accountants? [15].

• *Business*: Some have claimed that Blockchain will revolutionize business and redefine companies and economies. BC protocols facilitate businesses to use new methods of processing digital transactions. BC technology has a large potential to transform business and to bring significant efficiencies to global supply chains, payment system, remittances, national digital currency, financial transactions, banking, asset ledgers, insurance, real estate, stock exchange, etc. The distributed ledger technology systems enable businesses and banks to streamline internal operations, dramatically reducing mistakes caused by traditional methods for reconciliation of records. For example, the transfer of a share of stock can now take up to a week, but with BC it could happen in seconds [16].

• *Healthcare*: Appling BC in healthcare serves to improve patient care. BC technology offers patients and care-givers the ability to securely share patient identity and healthcare information across platforms. Imagine a future where patients hold the keys to their healthcare passport. Imagine a better quality of care for both patients and care providers [17].

• *Supply Chain*: This is one of the most obvious Blockchain application. Different partners can access the distributed ledger with the necessary permissions. BC can be used to capture information about the shipment of goods. Each participant in a supply chain can see the movement of goods through the supply chain and understand where a particular container is in transit.

• *Internet of Things*: IoT will play a major role in both

civilian and military contexts. IoT security is a serious issue. BC technology can be used to enhance security of IoT. It has the potential to facilitate secure sharing of IOT datasets and securing IoT systems [18]. BC integrates and automates machine-to-machine and IoT payment network for the machine economy.

- *Notary Public*: BC can be used to verify authenticity of a document without the need of centralized authority. Using BC for notarization secures the privacy of the document. It eliminates expensive notarization fees and ineffective means of transferring documents. Law firms are using BC technology as a cost effective way to certify documents [19].

- *e-Voting*: On regular basis, votes are recorded, counted, and checked by a central constituted authority. Blockchain-enabled e-voting enables voters to carry out these tasks by themselves and hold a copy of the voting record. An illegitimate vote will not be allowed because other voters will notice that it is not in agreement with the rules. This approach may work well for a minor organizational elections with a small number of voters and limited resources [20].

- *Copyright and Royalties*: These are a big issue in creative sectors like music, films, books, etc. The BC technology is quite important in ensuring security and transparency in the creative industries. There are many instances where music, films, art, book, etc. is plagiarized and due credit is not given to the original artists. This can be rectified using Blockchain which has a detailed ledger of artist rights. The payment of royalties can also be managed using digital currencies like Bitcoin [21].

Blockchain can also be used in identity management, utilities, real estate, law, tax reporting, postal service, microgrid, wireless networks, agriculture, architecture, patents, aircraft maintenance, online gaming, food safety, and music industry. These applications are illustrated in Figure 1.8 [22].

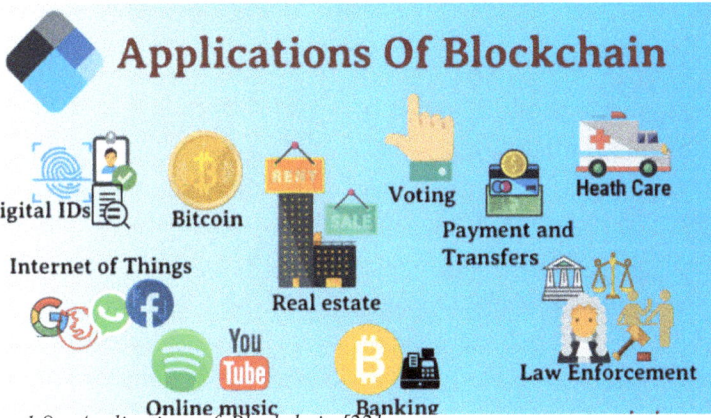

Figure 1.8   Applications of Blockchain [22].

Some of these applications will be discussed fully later in the book. The potential applications for BC technology are almost without limit.

1.5 BENEFITS

The potential benefits of BC extend into business, political, humanitarian, social, technological, and scientific realms. There are a couple of reasons why so many people in the technology and financial sectors are excited about the promise BC holds. By allowing digital information to be distributed but not copied,   BC technology has created the backbone of a new form of Internet. There is no single point of failure from which digital assets can be hacked or corrupted. Other benefits include the following:

- *Trust*: Blockchain is a great solution to the age-old human problem of trust. It enables trustless networks by allowing parties to conduct transactions even though they do not trust each other. The absence of a trusted middleman results in faster reconciliation between parties. BC removes the intermediary and moves towards democratization and decentralization [23].

- *Efficient Transactions*: BC transactions can be completed in minutes. This is particularly useful for cross-border trades, which usually take much longer because of time zone issues and the fact that all parties must confirm payment processing.

- *Decentralization*: Blockchain is a decentralized system. This means it eliminates the necessity of a central administrator and dependency on a third-party organization, where they may cause a lot of data security issues. The decentralized nature of BCs makes them an equality technology that can be used to expand freedom, actualization, and realization of all entities, both human and machine.

- *Transparency*: This implies that anyone can see and verify the transactions on a blockchain via the Internet. Because of the decentralized nature of the Bitcoin blockchain, all transactions can be transparently viewed by either having a personal node or using blockchain explorers that allow anyone to see transactions occurring live.

- *Traceability*: Information in a blockchain is organized chronologically. This makes it easy to track a specific transaction and trace it back to its source.

- *Integrity*: Data records are verified by consensus and available to all participants in the blockchain. They cannot be changed or deleted, which ensures data integrity.

- *Security*: Blockchain is a revolutionary technology because it helps reduce security risks, stamp out fraud, and bring transparency in a scalable way. Data can only be added to a Blockchain. Blockchain is designed to provide many security attributes, such as, consistency, tamper-proof, pseudonymity, resistance to double-spending and Distributed Denial-of-Service (DDoS) attacks.

- *Scalability*: As the number of users in a blockchain grows, so does the number of operations. The computational power required for these operations may outpace the workload that hard disks are realistically able to handle.

- *Speed*: Private BC transactions occur at greater speed as compared to public blockchains. That means the transactions per second (TPS) rate is higher in the case of private blockchains. This is because there is a limited number of nodes in a private network as opposed to a public network.

- *Reducing Costs*: BC technology can help organizations save money. It increases the efficiency of transaction processing. The ability of blockchain to streamline clearing and settlement directly translates into cost savings in the process. As a result, blockchain development firms can assist businesses in saving money by removing the middlemen.

These benefits make some to believe that BC has become the fifth disruptive computing paradigm after mainframes, PCs, the Internet, and mobile/social networking [24].

1.6 CHALLENGES

In spite of all its potential and benefits, Blockchain is not a silver bullet. Blockchain technology is not without challenges. There are hurdles to commercial adoption. Blockchain still faces some challenges such as scalability, energy consumption, consumer privacy, and legal considerations. These challenges are discussed as follows:

- *Security*: A major challenge is security. Companies need to have a security standards and systems to protect them from attackers or bad actors. In spite of these challenges, the demand for Blockchain-based services is on the rise and the technology is advancing at a rapid pace. Developers have built an array of applications on Blockchains.

- *Computational Power*: Managing the Blockchains requires substantial computational power in order to maintain security. The process of proof-of-work is highly energy consuming as it needs specialized systems to run a special algorithm.

- *Regulations*: Many in the crypto space have expressed concerns about government regulation over cryptocurrencies. Regulating and standardizing digital currency and money transmission is difficult. Engineers who manage BC projects need to follow the rules imposed by authorities, regulations, and standards.

- *Legal Challenges*: There are legal challenges surrounding Blockchain. Blockchain will disrupt all kinds of legal work, notary publics, contracts, lawyers, and judges.

- *Complexity*: The ever-growing size of the Blockchain is considered by some to be a problem, creating issues of storage and complexity.

1.7 CONCLUSION

Blockchain technology has emerged as a revolutionary concept that has the capability to transform various industries. Its main characteristics, such as decentralized ledgers and transparency of transactions, offer opportunities for increased efficiency, security, and trust in digital transactions. Blockchain is a member of the larger family of distributed-ledger technologies, which encompass all techniques for decentralized record keeping of transactions and data sharing across multiple servers, institutions, or nations. It is perhaps the main technological innovation of Bitcoin. It is the invisible technology that is disrupting the world. It is the heart of the fourth industrial revolution.

Recognizing BC as a revolutionizing technology across the industries, many people are excited about the possibilities that Blockchain technology will bring. Financial managers understand that the Blockchain has the potential to change the financial world. The job market looks promising for blockchain enthusiasts and it is expected to grow exponentially in the coming years. Figure 1.9 shows different job roles and opportunities for Blockchain experts [25].

Figure 1.8   Applications of Blockchain [22].

For more information on Blockchain, one should consult books in [16,24,26-34] and the following journal related to it: Blockchain: Research and Applications.

REFERENCES

[1] A. Guadamuz and C. Marsden, "Blockchains and Bitcoin: Regulatory responses to cryptocurrencies," Peer-reviewed Journal on the Internet, vol. 20, no. 12, Dec. 2015.

[2] M. N. O. Sadiku, Y. Wang, S. Cui, and S. M. Musa, "A primer on Blockchain," International Journal of Advances in Scientific Research and Engineering, vol. 4, no. 2, February 2018, pp. 40-44.

[3] C. M. M. Kotteti and M. N. O. Sadiku, "Blockchain technology," International Journal of Trend in Research and Development, vol. 10, no. 3, May-June 2023, pp. 274-276.

[4] "Blockchain," Wikipedia, the free encyclopedia

https://en.wikipedia.org/wiki/Blockchain

[5] S. Nakamoto, "Bitcoin: a peer-to-peer electronic cash system,"

https://bitcoin.org/bitcoin.pdf

[6] "The beginning of a new era in technology: Blockchain traceability,"

https://www.visiott.com/blog/blockchain-traceability/#:~:text=The%20Beginning%20of%20a%20New,money%20without%20a%20central%20bank.

[7] "The CIO's guide to Blockchain,"

https://www.gartner.com/smarterwithgartner/the-cios-guide-to-blockchain#:~:text=True%20blockchain%20has%20five%20elements,%2C%20immutability%2C%20tokenization%20and%20decentralization.

[8] "Blockchain structure,"

https://www.geeksforgeeks.org/blockchain-structure/

[9] G. O. R. Cruz, "What is Blockchain?" June 2022,

https://money.com/what-is-blockchain/

[10] M. Iansiti and K. R. Lakhani, "The truth about Blockchain," Harvard Business Review, Jan./Feb. 2017.

https://hbr.org/2017/01/the-truth-about-blockchain

[11] W. T. Tsai et al., "A system view of financial blockchains," Proceedings of IEEE Symposium on Service-Oriented System

Engineering, 2016, pp. 450-457.

[12] S. P. Mohanty et al., "PUFchain: Hardware-assisted blockchain for sustainable simultaneous device and data security in the Internet of everything (IoE)," IEEE Consumer Electronics Magazine, vol. 9, no. 2, March 2020, pp. 8-16.

[13] "Blockchain 4.0," December 2022,

https://www.bbvaopenmind.com/en/technology/digital-world/blockchain-4-0/

[14] "U.S. companies and their cryptocurrency holdings," May 2022,

https://www.reuters.com/business/finance/us-companies-their-cryptocurrency-holdings-2022-05-12/

[15] V. Shermin, "Disrupting governance with blockchains and smart contracts," Strategic Change, vol. 26, no. 5, 2017, pp. 511-522.

[16] M. Gupta, Blockchain for Dummies. Hoboken, NJ: John wiley & Sons, 2017.

[17] S. Manski, "Building the blockchain world: technological commonwealth or just more of the same?" Strategic Change, vol. 26, no. 5, 2017, pp. 511-522.

[18] M. Banerjee, J. Lee, and K. K. R. Choo, "A blockchain future to Internet of things security: a position paper," to appear in Digital Communication and Networks, 2017.

[19] M. Crosby et al., "BlockChain Technology", Sutardja Center for Entrepreneurship & Technology Technical Report, UC Berkeley, Oct. 2015.

[20] P. Boucher, "How blockchain technology could change our lives," European Parliamentary Research Service, Feb. 2017.

[21] "Top applications of blockchain in the real world,"

https://www.geeksforgeeks.org/how-hotstar-and-jiocinema-manages-millions-of-live-users/?ref=rightsidebar-card

[22] P. Pedamkar, "Applications of Blockchain," June 2023,

https://www.educba.com/applications-of-blockchain/

[23] K. Christidis and M. Devetsikiotis, "Blockchains and smart

contracts for the Internet of things," IEEE Access, vol. 4, 2016, pp. 2292-2303.

[24] M. Swan, Blockchain: Blueprint for a New Economy. Sebastopol, CA: O'Reilly Media, 2015.

[25] "Top 10 reasons why you should learn Blockchain," January 2023,

https://www.edureka.co/blog/top-10-reasons-to-learn-blockchain/

[26] T. Laurence, Blockchain For Dummies. Wiley, 3rd edition, 2023.

[27] T. Laurence, Introduction to Blockchain Technology: The Many Faces of Blockchain Technology in the 21st Century. Van Haren Publishing, 2019.

[28] Z. Usmani, Introduction to Blockchain with Case Studies. Gufhtugu Publishers, 2018.

[29] A. Lewis, The Basics of Bitcoins and Blockchains: An Introduction to Cryptocurrencies and the Technology that Powers Them. Mango, 2018.

[30] R. J. Robinson III, Introduction To Embedded Blockchain Cyber Security. Kindle Edition, 2018.

[31] A. Banafa, Introduction to Blockchain Technology. ʃRiver Publishers, 2023.

[32] T. Gaitatzis, A Programmer's Introduction to Blockchain: An Technical Explanation of How Blockchain Databases Work. BackupBrain Press, 2022.

[33] D. Drescher, Blockchain Basics: A Non-Technical Introduction in 25 Steps. Apress, 2017.

[34] Mastering Blockchain: A Deep Dive Into Distributed Ledgers, Consensus Protocols, Smart Contracts, Dapps, Cryptocurrencies, Ethereum, and More. Packt Publishing, 3rd edition, 2020.

CHAPTER 2

BLOCKCHAIN IN BUSINESS

*"I believe that Blockchain will do for trusted transactions what the
Internet has done for information."*

— Ginni Rometty

2.1 INTRODUCTION

The business world is constantly evolving, and technological advancements play a pivotal role in shaping its landscape. Traditional business models are no longer sufficient to tackle global competition due their inherent limitations. New developments are continually hitting the business world in recent years. Digital transformation allows business to take strategic decisions and leverage disruptive technologies in the marketplace. Blockchain is one of the major technologies driving business transformation. It is based on a peer-to-peer topology that ensures transactions' transparency, data resiliency, and security while reducing additional costs in running business operations [1].

In essence, Blockchain is a massive, decentralized ledger of transactions maintained by many different, decentralized sources. It may be regarded as the chain of blocks that contain data. It is a continuously updated digital record of who holds what. Information about transaction (the time, date, dollar value, and the participants) is encrypted into a "block" that is linked to other blocks to form a chain. From farmers to grocery store, each transaction is recorded on a Blockchain. Everyone in a Blockchain has access to the same information and there is no one central authority, providing transparency and continuous reconciliation [2].

Block chain is a disruptive technology that brings transformation in business processes. The main reason is to reap major benefits like transparency, decentralization, immutability, and security at reduced costs. A Blockchain creates a trusted record using cryptography. Blockchain technology is becoming increasing important in numerous industries such as healthcare, logistics, manufacturing, and business.Blockchain is valuable in business for entities transacting with one another. Across industries around the world, Blockchain is helping transform business [3].

This chapter provides an introduction to the use of Blockchain technology in business. It begins with presenting an overview on Blockchain to make the chapter self-contained. It covers some applications of BC in business. It highlights the benefits and challenges of BC in business. The last section concludes with comments.

2.2 OVERVIEW OF BLOCKCHAIN

The term "Blockchain" refers to the way BC stores transaction data – in "blocks" that are linked together to form a "chain." The chain grows as the number of transactions increases. Since every entry is stored as a block on a chain, the care you receive is added to your personal ledger. Figure 2.1 displays Blockchain architecture [4].

The first Blockchain was conceived in 2008 by an anonymous person or group known as Satoshi Nakamoto, who published a white paper introducing the concept of a peer-to-peer electronic cash system he called Bitcoin.

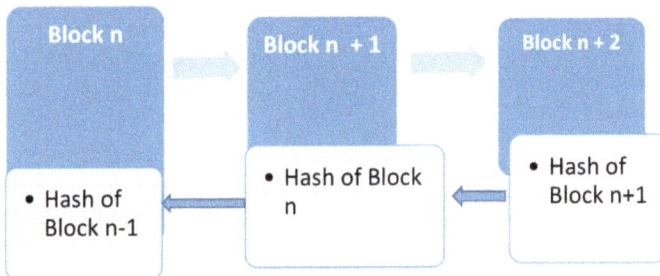

Figure 2.1   Blockchain architecture [4].

At its core, Blockchain is a distributed system recording and storing transaction records. In a Blockchain system, there is no central authority.   Instead, transaction records are stored and distributed across all network participants. Rather than having a centrally located database that manages records, the database is distributed to the networks and transactions are kept secure via cryptography. BC eliminates the need for a middleman that traditionally may facilitate such transactions [5].

Fundamentally, Blockchains are distributed digital database that record and maintain a list of transactions taking place in real time. They may also be regarded as decentralized ledgers that sequentially record transactions or interactions among users within a distributed network. They have the following properties [6]:

- Firstly, they are autonomous. They run on their own, without any person or company in charge.

- Secondly, they are permanent. They are like global computers with 100 percent uptime. Because the contents of the database are copied across thousands of computers, if 99 per cent of the computers running it were taken offline, the records would remain accessible and the network could rebuild itself.

- Thirdly, they are secure and tamper-proof. Each record in Blockchain is time stamped and stored cryptographically. The encryption used on Blockchains like Bitcoin and Ethereum is industry standard, open source, and has never been broken.

- Fourthly, they are open, allowing anyone to develop products and services on them.

- Fifthly, as Blockchain is a shared system, costs are also shared between all of its users.

The Blockchain was designed so transactions are immutable, i.e. they cannot be deleted. Thus, Blockchains are secure and meddle-free by design. Data can be distributed, but not copied. When it comes to digital assets and transactions, you can put almost anything on a Blockchain. Different scenarios call for different Blockchains. Blockchain is used for different purposes as depicted in Figure 2.2 [7].

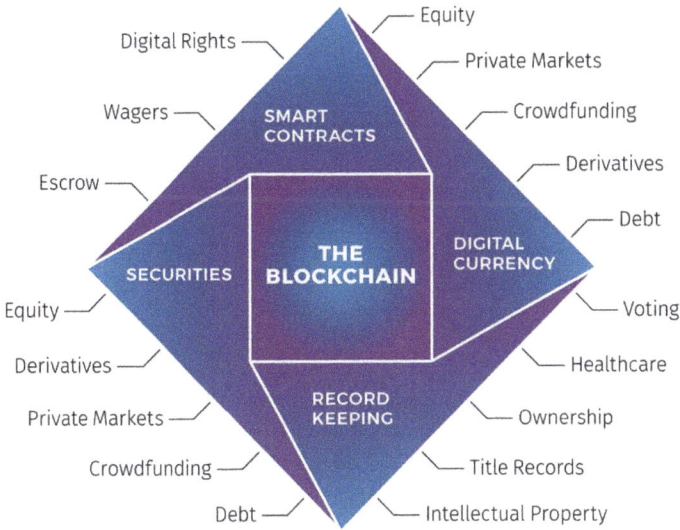

Figure 2.2 Different purposes of Blockchain [7].

The BC technology currently has the following features [8,9]:

1. *Peer-to-Peer (P2P) Network*: The first requirement of BC is a network, an infrastructure shared by multiple parties. This can be a LAN at a small scale or the Internet at a large scale. All nodes participating in a BC are connected in a decentralized P2P network. Transactions are broadcast to the P2P network. Due to some limitations of P2P networks, some vendors have provided cloud-based BCs.

2. *Cascaded Encryption*: A BC uses encryption to protect transaction data. Blocks are encrypted in a cascaded manner, i.e. the encryption result of the previous block is used in encrypting the current block. The BC is secured by public key cryptography, with each peer generating its own public-private key pairs.

3. *Distributed Database*: A BC is digitally distributed across a number of computers. Each party on a BC has access to the entire database and no single party controls the data or the information. Since BC is decentralized, there is no need for central authorizes such as banks.

4. *Transparency with Pseudonymity*: Each node or participant on a Blockchain has a unique 30-plus-character alphanumeric address that identifies it. Users can choose to remain anonymous or provide proof of their identity to others.

5. *Irreversibility of Records*: Once a transaction is entered in the database and the accounts are updated, the records cannot be altered. Records on the database is permanent, chronologically ordered, and available to all others on the network.

2.3 APPLICATIONS OF BLOCKCHAIN IN BUSINESS

The applications of Blockchain are no longer the pie-in-the-sky dreams. Although the most popular application is probably cryptocurrency, experts claim that Blockchain can be used for everything from data management to regulatory compliance. This technology has the potential to disrupt nearly every industry and solve challenges facing any business. More and more industry analysts agree that Blockchain technology would have a significant impact on businesses, including the way they are financed and run, how they generate value, and how they conduct basic functions including marketing, accounting, and motivating employees and customers [10]. Some of these business applications of Blockchain are displayed in Figure 2.3 [11].

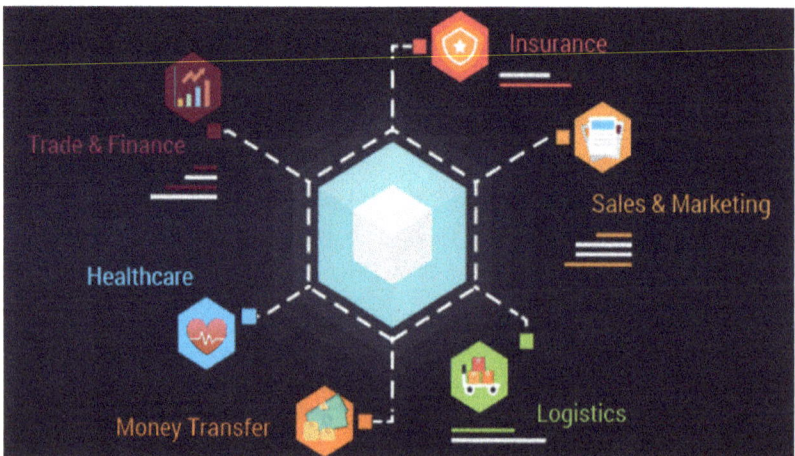

Figure 2.3   Some business applications of Blockchain [11].

The applications include the following [1,12-14]:

- *Smart Contracts*: The foremost application of Blockchain in business is smart contracts, which are essentially a kind of self-executing contract where all the terms and conditions from both parties are written in the form of codes. Smart contracts enable a way for organizations to automatically handle large amounts of transactions. They are executed in an automated way and do not need human intervention or paper trails. They are decentralized and secured. They can be divided into two broad categories: smart legal contracts and code-based. Lawyers are using Blockchain to create smart contracts.

- *Human Resources*: Blockchain plays a major role in the recruitment process. It aids organizations by saving time required for verifying all the documents and hiring the potential candidate.

- *Marketing*: Blockchain technology improves marketing campaigns. It empowers marketers to keep a real-time track of client information and customer behavior.

- *Customer Engagement*: Blockchain opens new doors for engaging a wider target audience. The combination of Blockchain and customer engagement brings provides some advantages.

- *Supply Chain Management*: Supply chains are the organizational and logistical systems by which goods are transported from a factory to the consumer. Blockchain can be used to track goods and materials throughout the supply chain of a manufacturing company. The technology enables companies to track their products/services from manufacturing to transportation, and delivery at the consumer end in real time. Blockchain eliminates supply chain disputes because every supplier and producer can view the chain of ownership on the same ledger. The transparency of Blockchains has benefits in supply chain management, visibility, and traceability.

- *Financial Services*: This industry is beginning to use Blockchain to develop new services and save on costs.

Blockchain is used mainly in finance for cryptocurrency and decentralized finance. Cryptocurrency is gaining credibility as an alternative payment used by government central banks and big payment providers such as Mastercard and Visa. Blockchain has its own secure system that is capable of keeping track of the various multi-million dollar transactions that are carried out daily in the banking industry. For example, using Blockchain ICBC (the largest bank of China) intends to eliminate the possibility of forgery of various documents. Since money is becoming digital, trading cryptocurrency is a profitable business concept.

- *Money Transfer*: Today, the banking sector faces various kinds of problems, while Blockchain technology has a high potential for solving them. Blockchain has made seamless, immediate, and affordable cross-border payments possible. This technology suits the financial transaction industry because it can provide secure and immutable operations while keeping sensitive client data intact. Money transfers come with low fees and can be completed in seconds from and to any part of the world.

- *Real Estate*: The real estate industry involves complex processes, extensive paperwork, and the need for intermediaries. Blockchain technology can simplify these processes by digitizing property records, titles, and contracts. This digitization improves transparency, reduces fraud, and minimizes the involvement of intermediaries, resulting in cost savings for buyers and sellers.

- *Government*: Blockchain could make it easier to transmit personal information electronically, vote online, acquire a passport, and prepare legal documents. As with banks, governments' key record-keeping and verifying functions can be enabled by Blockchain infrastructure to achieve large administrative savings.

- *Voting*: Blockchain can be used to create secure and transparent voting systems, reducing the chances of election fraud. Votes can be moved along a Blockchain in a neutral, accurate, and secure way. Using Blockchain alters modern notions of democracy and strengthens the validity of election results. Blockchain's immutability and transparency are both necessary for successful elections. In 2018, USA and Sierra Leone in West Africa used an innovative Blockchain-based voting system from start-up Agora. The voting system made it possible to make the elections transparent and cheaper and caused the citizens to be satisfied with the electron process.

- *Healthcare*: Blockchain could be a solution to the woes of beleaguered industries like healthcare. Medical errors are the third leading cause of death in the United States due to communications challenges between providers. Traditional data-sharing methods leave patient records vulnerable to theft. Blockchain technology eliminates this risk by creating secure "blocks" of data. Blockchain could be the key to unlocking the value of data availability and exchange across providers, patients, insurers, and researchers. Smart contracts could give patients more control over their data. Figure 2.4 shows a typical use of Blockchain in healthcare [15].

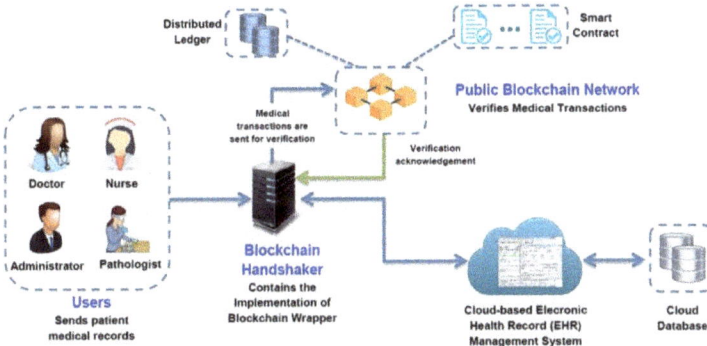

Figure 2.4   A typical use of Blockchain in healthcare [15].

- *Cybersecurity*: Blockchain's strong encryption and other security safeguards make it another tool in the security toolbox. Cybersecurity includes the protection of confidential intellectual property. Using the Blockchain in the intellectual property registry will help authors to get maximum information about copyright on their content. Cybersecurity also includes the protection of confidential intellectual property. Blockchain in cybersecurity is displayed in Figure 2.5 [16].

BLOCKCHAIN IN CYBERSECURITY

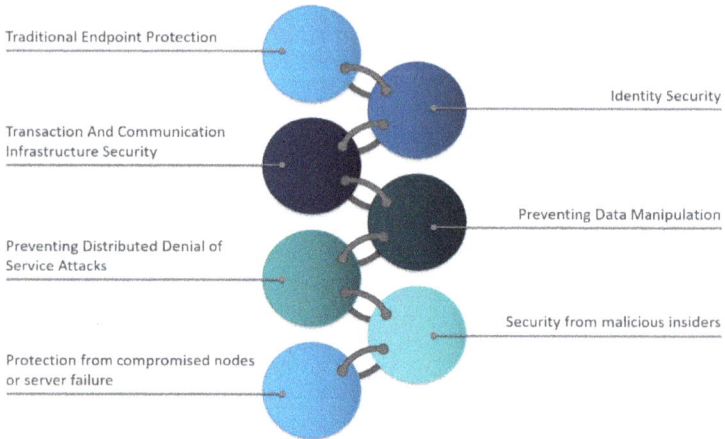

Traditional Endpoint Protection

Identity Security

Transaction And Communication
Infrastructure Security

Preventing Data Manipulation

Preventing Distributed Denial of
Service Attacks

Security from malicious insiders

Protection from compromised nodes
or server failure

Figure 2.5   Blockchain in cybersecurity [16].

- *Digital Identity*: Security is a top priority for small businesses. Blockchain can be used for secure identity verification, reducing the risk of identity theft and fraud. Individuals have more control over their personal information and who can access it. The use of Blockchain in business for digital identity facilitates users to protect and maintain their identity and see how they can access their information and use it for any purpose.

- *Insurance*: Blockchain can automate and expedite the claims process by providing a transparent and tamper-proof record of events. Insurers can use Blockchain to assess risks more accurately by analyzing vast amounts of data from various sources.

- *Energy Sector*: The energy sector is undergoing a transformation towards renewable energy sources and decentralized energy systems. Blockchain can play a vital role in this transition by facilitating peer-to-peer energy trading and tracking renewable energy generation and consumption.

- *Education*: This is one of the most desirable and important industries in the global community. A Blockchain can be used to store sensitive data, such as coursework. Thus, no one will be able to take someone else's work for their work. The grades recorded in the Blockchain cannot be changed or erased, which guarantees honesty. Today, the main method of checking a diploma is to send a request to an educational institution, which sometimes takes about a month, and therefore requires too much time. The Blockchain will provide this data in seconds and speed up the hiring process. Figure 2.6 shows the uses of BC in education [16].

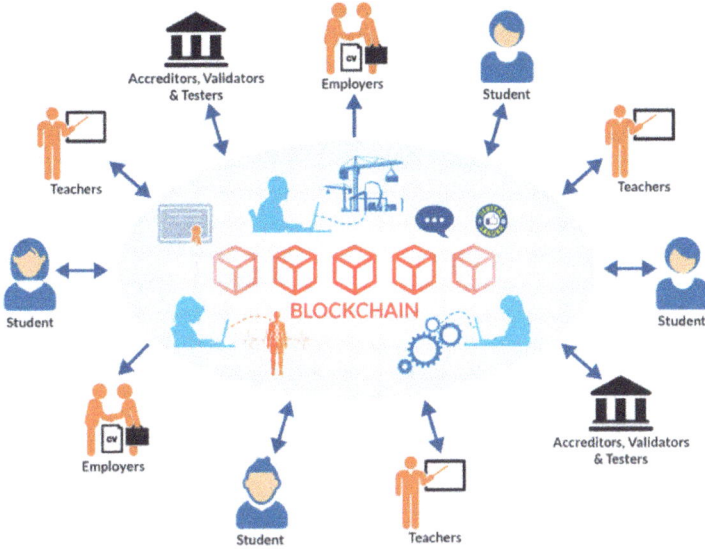

Figure 2.6   Uses of Blockchain in education [16].

- *Retail*: Technological advancements help retail businesses in countless ways. Figure 2.7 illustrates the uses of Blockchain in retail [17].

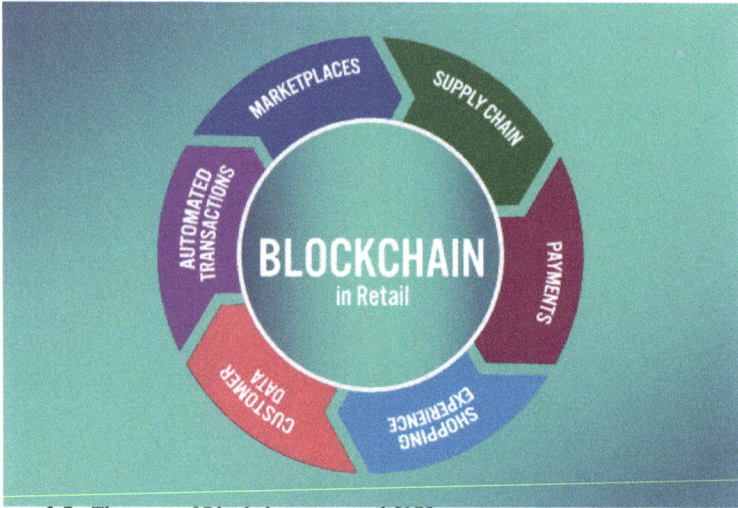

Figure 2.7   The uses of Blockchain in retail [17].

Some businesses have started to use Blockchain in various of retail companies. These include Walmart, Carrefour, Amazon, Alibaba, and Home Depot. For example, Walmart uses Blockchain for security reasons. The use of Blockchain technology has helped Walmart to reduce from one week to a few seconds the time needed "to trace the source of sliced mangoes." As shown in Figure 2.8, Amazon uses its Amazon Web services platform to provide Blockchain solutions for companies [17].

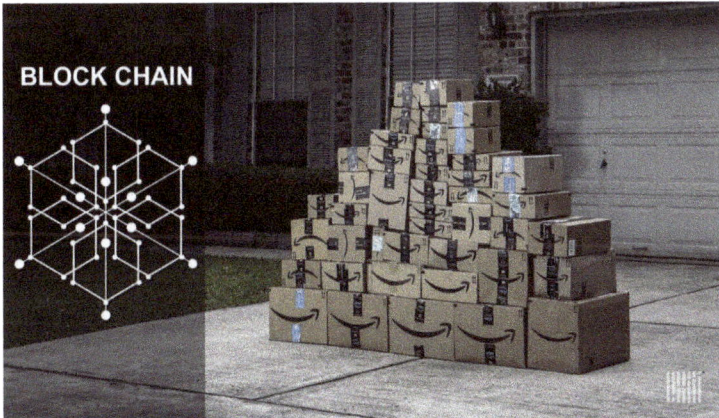

Figure 2.8 Amazon provides Blockchain solutions for companies [17].

2.4 BENEFITS

The key strength of Blockchain is its decentralized nature. Blockchain technology has the potential to transform the way businesses operate by offering numerous benefits. Because Blockchain removes middlemen and automates processes, it has the potential to save businesses costs, speed up e-commerce, and enable new lines of business. It also has a significant role in lowering carbon emissions. Implementation of Blockchain technology in business environment establishes transparency, decentralizes processes, improves security, achieves scalability, increases accountability, reducing investment costs, etc. These benefits are further explained as follows [18]:

- *Decentralized Process*: Every participant in the network owns their data, can access the information history, and confirm new transactions. This is helpful when businesses must interact as peers but no one wants responsibility for maintaining the system.

- *Transparency*: The lack of transparency along global supply chains poses challenges in the areas of fraud, pollution, human rights abuses, intellectual property theft, and inefficiencies. The Blockchain is transparent and unchangeable. The transactions occur in consensus with the participating parties. It is convenient, trustworthy, and transparent. Blockchain benefits derive from the trust it fosters, its built-in privacy, and its transparency. The key properties of integrity, resilience, and transparency of Blockchain make it an attractive option to enterprises to revolutionize their business processes.

- *Digital and Automated Records*: The transactions are digital and automated, thereby reducing paper-based transactions.

- *Automated Validation and Verification*: Blockchain utilizes consensus protocols to validate transactions and does not rely on a single point of authority.

- *Improved Efficiency and Speed*: Blockchain transactions do not need intermediaries leading to a faster turnaround.

- *Enhanced Security*: A major role of Blockchain in business is to introduce robust security measures in the environment. Blockchain is highly secure due to its digital signature and encryption. The data is stored in a number of places and immune to hackers. Since Blockchains are decentralized, there are no centralized servers or databases. This implies that there are no single points of failure.

- *Trust*: Trust makes it possible to do business with unknown parties. Blockchain spreads trust everywhere. Whether it is between people or organizations, relationships flourish when there is more trust. Blockchain encourages trust between entities where trust is either lacking or unproven.

- *Cost Reduction*: Cost management and reduction are benefits of Blockchain technology because Blockchain technology can also help organizations save money. The value of Blockchain is in driving cost reduction to enabling entirely new business models and revenue streams. It eliminates cost for mediators or intermediaries.

- *Intellectual Property Protection*: Protecting intellectual property rights is crucial for individuals and businesses. Artists, musicians, and content creators can use Blockchain to manage their intellectual property rights, ensuring fair compensation, and protecting against piracy.

- *Scalability*: This refers to the capability of a business to perform well despite increased workload. Ethereum is the most widely used network for smart contracts, and its transition to proof-of-stake further improves scalability and security.

2.5 CHALLENGES

Despite its potential and benefits, we are still in the early days of the Blockchain. Businesses that implement Blockchain come across various challenges that demand different solutions. The challenges include [2,19]:

- *Energy Use*: Verifying transactions to add to the Blockchain is energy-intensive because of the computer power required to do all the computations.

- Processing Speed: Because of all the computations involved, the processing time is slow.

- *Interoperability*: Ensuring different Blockchain networks can work together seamlessly is crucial for widespread adoption. Currently there are many Blockchains that do not interact with each other. Blockchain has the difficulty of resolving the "coopetition" paradox to establish common standards.

- *Lack of Standards*: The lack of common standards and clear regulations is a major limitation on Blockchain. Standards can be established with relative ease if there is a single dominant player or a government agency that can mandate the legal standing. Leaders should act now to maintain their competitive edge and take advantage of the opportunity to set industry standards.

- *Lack of Adoption*: Blockchain, especially public Blockchain, requires a large number of participants for most of its benefits to be realized.

- *Cost*: The immaturity of Blockchain technology increases the switching costs.

- *Threat*: Many people view Blockchain as a threat to the current legacy of corporations and companies that require centralization. Since Blockchain technology is decentralized, it can be used by anybody who have been granted permission.

- *Security*: The first priority for any financial company is in the area of security. Blockchain transactions can never be faked or altered, making the technology a secure foundation to build a business on. While Blockchain technology itself is secure, the applications built on it might have vulnerabilities.

- *Privacy*: Ensuring the privacy of sensitive data can be a challenge using Blockchain. Striking the right balance between transparency and privacy is crucial.

The next generation of Blockchain technology will address these challenges. The issues need to be resolved before BC goes mainstream.

2.6 CONCLUSION

Simply put, a Blockchain is a database or ledger that is distributed across a private or public computer network. It is an immutable decentralized way to securely store data in blocks that are linked to each other using cryptography. It is a revolutionary change in how things work. Although Blockchain is still a relatively novel enterprise technology, it has the potential to deliver significant change and transform many industries. It is a universal technology, which can be easily adopted in any field.

Blockchain seems to be changing everything in the business world. While Blockchain holds immense potential for revolutionizing various aspects of business operations, it is important for businesses to carefully assess their needs, understand the technology, and consider the challenges before implementing Blockchain solutions. There are many industries that are already utilizing Blockchain technology to their advantage. If you are interested in Blockchain, you should consider becoming a certified Blockchain engineer. More information about Blockchain in business can be found in the books in [20-35].

REFERENCES

[1] "10 Ways to embrace Blockchain for business transformation," December 2022,

https://appinventiv.com/blog/importance-of-Blockchain-for-business/

[2] "What is Blockchain and what does it mean for your business?

https://www.bdc.ca/en/articles-tools/blog/what-is-Blockchain-what-does-it-mean-for-your-business

[3] M. N. O. Sadiku, U. C. Chukwu, and J. O. Sadiku, "Blockchain in business," European Journal of Business Startups and Open Society, vol. 3, no.4, April 2023, pp. 12-20.

[4] A. Pal, C. K. Tiwari, and N. Haldar, "Blockchain for business management: Applications, challenges and potentials," Journal of High Technology Management Research, vol. 32, 2021.

[5] M. N. O. Sadiku, Y. Wang, S. Cui, and S. M. Musa, "A primer on Blockchain," International Journal of Advances in Scientific Research and Engineering, vol. 4, no. 2, February 2018, pp. 40-44.

[6] S. Depolo, "Why you should care about Blockchains: the non-financial uses of Blockchain technology," March 2016,

https://www.nesta.org.uk/blog/why-you-should-care-about-Blockchains-non-financial-uses-Blockchain-technology

[7] "Benefits of Blockchain: A business sector perspective,"

https://www.e-zigurat.com/innovation-school/blog/benefits-of-Blockchain/

[8] M. Iansiti and K. R. Lakhani, "The truth about Blockchain," Harvard Business Review, Jan./Feb. 2017.

https://hbr.org/2017/01/the-truth-about-Blockchain

[9] W. T. Tsai et al., "A system view of financial Blockchains," Proceedings of IEEE Symposium on Service-Oriented System Engineering, 2016, pp. 450-457.

[10] P. Gatomatis, K. Tsiomos, and N. Bogonikolos, "Using Blockchain for business & marketing improvement," Journal of Marketing Development and Competitiveness, vol. 15, no. 2, 2021.

[11] A. Shah, "Implementing Blockchain technology in business,"
https://www.fusioninformatics.com/blog/implementing-Blockchain-technology-business/

[12]A. Uzialko, "Beyond Bitcoin: How Blockchain is improving business operations," December 2017,

https://7wdata.be/data-analysis/beyond-bitcoin-how-Blockchain-is-improving-business-operations/

[13] D. Essex, "Blockchain for businesses: The ultimate enterprise guide," June 2021,

https://www.techtarget.com/searchcio/Blockchain-for-businesses-The-ultimate-enterprise-guide

[14] "The impact of Blockchain on the business world," May 2023,

https://www.linkedin.com/pulse/impact-Blockchain-business-world-virtual-height-it-services-pvt-ltd

[15] "21 Blockchain business ideas with huge potential for your startup," March 2023,

https://sumatosoft.com/blog/Blockchain-business-ideas-for-your-startup

[16] E. Tarasenko, "Top 9 Blockchain business opportunities," February 2023,

https://merehead.com/blog/top-9-Blockchain-business-opportunities/

[17] B. Brown, "10 Retail companies using Blockchain technology," September 2021,

https://www.getdor.com/blog/2021/09/14/retail-companies-using-Blockchain-technology/ [18] K. Adams, " Benefits of Blockchain technology for businesses," August 2022,

https://www.datasciencecentral.com/benefits-of-Blockchain-technology-for-businesses/

[19] "Blockchain beyond the hype: What is the strategic business value?" June 2018, https://www.linux.com/news/blockchain-beyond-hype-what-strategic-business-value/

[20] H. K. Baker, E. Nikbakht, and S. S. Smith, The Emerald Handbook of Blockchain for Business. Emerald Publishing, 2021

[21] J. S. Arun, J. Cuomo, and N. Gaur, Blockchain for Business. Addison-Wesley Professional, 2019.

[22] C. Bai, J. Cordeiro, and J. Sarkis (eds.), Blockchain Technology: Business, Strategy, the Environment, and Sustainability. Wiley, 2022.

[23] S. S. Tyagi and S. Bhatia (eds.), Blockchain for Business: How it Works and Creates Value. Wiley, 2021.

[24] A. Tapscott and D. Tapscott, Blockchain Revolution: How the Technology Behind Bitcoin Is Changing Money, Business, and The. Portfolio, 2018.

[25] W. Mougayar, Business Blockchain, The MP3 CD. Audible Studios on Brilliance Audio; Unabridged edition, 2017.

[26] W. Mougayar, The Business Blockchain: Promise, Practice, and Application of the Next Internet Technology. Wiley, 2016.

[27] J. Arun, J. Cuomo, and N. Gaur, Blockchain for Business. Addison-Wesley Professional, 2019.

[28] Y. Kalfoglou, Blockchain for Business: A Practical Guide for the Next Frontier. Taylor & Francis, 2021.

[29] D. Furlonger and C. Uzureau, The Real Business of Blockchain: How Leaders Can Create Value in a New Digital Age. Harvard Business Review Press, 2019.

[30] M. Attaran, and A. Gunasekaran, Applications of Blockchain Technology in Business: Challenges and Opportunities. Springer, 2019.

[31] M. V. Rijmenam and P. Ryan, Blockchain: Transforming Your Business and Our World. Taylor & Francis, 2018.

[32] P. Lipovyanov, Blockchain for Business 2019: A User-friendly Introduction to Blockchain Technology and Its Business Applications. Packt Publishing, 2019.

[33] E. Nikbakht, H. K. Baker, and S. S. Smith, The Emerald Handbook of Blockchain for Business. Emerald Publishing Limited, 2021.

[34] Harvard Business Review et al., Blockchain: The Insights You Need from Harvard Business Review. Harvard Business Review Press, 2019.

[35] U. Hacioglu (ed.), Digital Business Strategies in Blockchain Ecosystems: Transformational Design and Future of Global Business. Springer, 2019.

CHAPTER 3

BLOCKCHAIN IN FINANCE

"The blockchain is a distributed network that solves all the problems that we have of finance, but more broadly, it's like a philosophy. It's a way of life."

— Mike Cernovich

3.1 INTRODUCTION

Finance industry is one of the largest booming sectors. This sector witnesses millions of transactions worth trillions of dollars daily. The global financial industry provides services to billions of people daily while managing trillions of cash. The financial industry is under immense pressure and is facing numerous challenges on a global basis. For centuries, the financial system has been heavily centralized. Along with other government entities, financial institutions joined hands to govern almost all transactions, from money issuance to lending and investing. The advent of Blockchain has changed the fundamental mechanics of how the financial system operates with its decentralized record-keeping feature. Blockchain is a digital database that enables simultaneous storage of certain operation records across numerous machines. Blockchain specialists claim that by bringing visibility and reducing friction along the lengthy list of transactions that typically precede financial interactions, Blockchain is enhancing security, reducing risk, and saving money [1].

Before Blockchain technology, people turned to gold or real estate when inflation hit its peak. Today, governments all over the globe have started opening up to Blockchain and crypto. The first to start this process was China with its "digital yuan." [2]. Blockchain is a burgeoning technology across a wide spectrum of industries. It is poised to shake up financial markets in a way that will benefit both consumers and financial institutions. A key reason is that its "core functions of verifying and transferring financial information and assets very closely align with Blockchain's core transformative impact" [3].

Blockchain (BC) is basically a ledger of recorded financial

transactions. This ledger is distributed, published, and stored in multiple locations. Blockchain is a decentralized record-keeping system that documents all transactions that happened on it. It facilitates safe, easy transactions, and builds trust between trading partners. When trust in the central hubs of finance is being increasingly questioned, decentralized systems like the Blockchain that reduce the need for such trust become attractive. Blockchain provides a potentially attractive alternative way to organize modern finance [4].

This chapter provides a primer on the use of Blockchain technology in finance. It begins with presenting an overview on Blockchain to make the chapter self-contained. It covers some applications of BC in finance. It highlights the benefits and challenges of BC in finance. The last section concludes with comments.

3.2 OVERVIEW OF BLOCKCHAIN

Blockchain (BC) technology is a permanent record of online transactions. It is a distributed tamper-proof database, shared, and maintained by multiple parties. It is a new enabling technology that is expected to revolutionize many industries, including business. It has the potential for addressing significant business issues. The BC technology allows participants to move data in real-time, without exposing the channels to theft, forgery, and malice.

The term "Blockchain" refers to the way BC stores transaction data – in "blocks" that are linked together to form a "chain." The chain grows as the number of transactions increases. Since every entry is stored as a block on a chain, the care you receive is added to your personal ledger. The first Blockchain was conceived in 2008 by an anonymous person or group known as Satoshi Nakamoto, who published a white paper introducing the concept of a peer-to-peer electronic cash system he called Bitcoin [5].

At its core, Blockchain is a distributed system recording and storing transaction records. In a Blockchain system, there is no central authority. Instead, transaction records are stored and distributed across all network participants. Rather than having a centrally located database that manages records, the database is distributed to the networks and transactions are kept secure via cryptography. BC eliminates the need for a middleman that traditionally may facilitate such transactions. Figure 3.1 shows how Blockchain works [6].

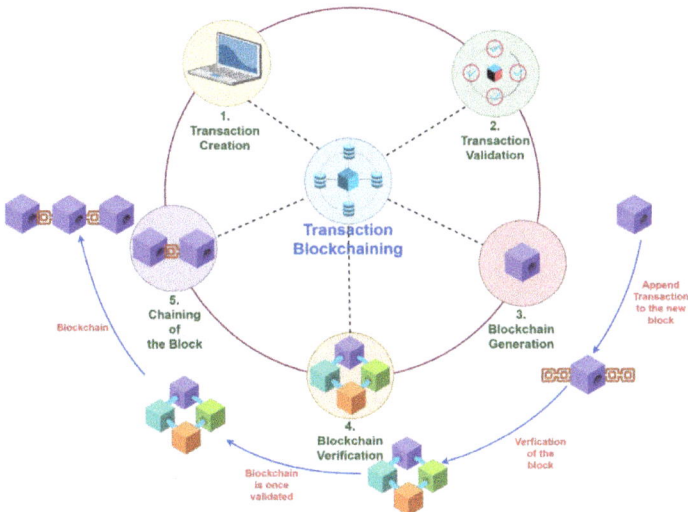

Figure 3.1   How Blockchain works [6].

Fundamentally, Blockchains are distributed digital database that record and maintain a list of transactions taking place in real time. They may also be regarded as decentralized ledgers that sequentially record transactions or interactions among users within a distributed network. They have the following properties [7]:

- Firstly, they are autonomous. They run on their own, without any person or company in charge.

- Secondly, they are permanent. They are like global computers with 100 percent uptime. Because the contents of the database are copied across thousands of computers, if 99 per cent of the computers running it were taken offline, the records would remain accessible and the network could rebuild itself.

- Thirdly, they are secure and tamper-proof. Each record in Blockchain is time stamped and stored cryptographically. The encryption used on Blockchains like Bitcoin and Ethereum is industry standard, open source, and has never been broken.

- Fourthly, they are open, allowing anyone to develop products and services on them.

- Fifthly, as Blockchain is a shared system, costs are also

shared between all of its users.

The Blockchain was designed so transactions are immutable, i.e. they cannot be deleted. Thus, Blockchains are secure and meddle-free by design. Data can be distributed, but not copied. When it comes to digital assets and transactions, you can put almost anything on a Blockchain. Different scenarios call for different Blockchains. Blockchain is used in different areas such as depicted in Figure 3.2 [8].

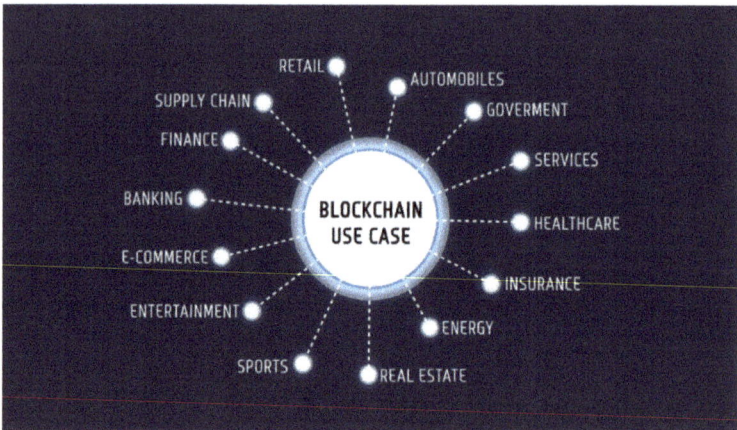

Figure 3.2   Different applications of Blockchain [8].

The BC technology currently has the following features [9,10]:

1. *Peer-to-Peer (P2P) Network*: The first requirement of BC is a network, an infrastructure shared by multiple parties. This can be a LAN at a small scale or the Internet at a large scale. All nodes participating in a BC are connected in a decentralized P2P network. Transactions are broadcast to the P2P network. Due to some limitations of P2P networks, some vendors have provided cloud-based BCs.

2. *Cascaded Encryption*: A BC uses encryption to protect transaction data. Blocks are encrypted in a cascaded manner, i.e. the encryption result of the previous block is used in encrypting the current block. The BC is secured by public key cryptography, with each peer generating its own public-private key pairs.

3. *Distributed Database*: A BC is digitally distributed across

a number of computers. Each party on a BC has access to the entire database and no single party controls the data or the information. Since BC is decentralized, there is no need for central authorizes such as banks.

4. *Transparency with Pseudonymity*: Each node or participant on a Blockchain has a unique 30-plus-character alphanumeric address that identifies it. Users can choose to remain anonymous or provide proof of their identity to others.

5. *Irreversibility of Records*: Once a transaction is entered in the database and the accounts are updated, the records cannot be altered. Records on the database is permanent, chronologically ordered, and available to all others on the network.

There are two types of Blockchains: public and private. Public Blockchains are cryptocurrencies such as Bitcoin, enabling peer-to-peer transactions. Private Blockchains use Blockchain-based platforms such as Ethereum or Blockchain-as-a-service (BaaS) platforms running on private cloud infrastructure. A private BC is an intranet, while a public BC is the Internet. Companies will be disrupted the most by public Blockchains.

3.3 APPLICATIONS OF BLOCKCHAIN IN FINANCE

The Blockchain is a decentralized, replicated, tamper resistant, append-only ledger of transactions. Blockchain technology has significantly impacted the financial industry by introducing new ways of conducting transactions, managing data, and ensuring security. Blockchain in financial services can offer multiple benefits and applications, which can help transform the finance industry. The applications include the following [11-15]:

- *Cryptocurrency*: Blockchain is the underlying technology for cryptocurrencies like Bitcoin and Ethereum. These digital currencies enable secure and direct peer-to-peer transactions without the need for intermediaries like banks. Cryptocurrency, also known as crypto-currency or crypto, is the most popular digital currencies. It received its name because it uses encryption to verify transactions. It is a form of decentralized digital currency that is not pegged to any fiat currency. It is the primary services of application of Blockchain technology that has achieved impressive results and made a real "monetary revolution." It is electronic money that banks and states do not regulate. It uses cryptography to secure and verify transactions in a network. When you use crypto as a form of payment, you also create a taxable event, which means you may owe capital gains taxes each time you purchase something with Bitcoin or Ethereum. Cryptocurrencies are regarded as virtual currencies because they are unregulated and exist only in digital form. There are thousands of cryptocurrencies. The two most popular cryptos are Bitcoin and Ethereum. The first cryptocurrency was Bitcoin, which was founded in 2009 and remains the best known today. Ethereum was developed in 2015 and is a Blockchain platform with its own cryptocurrency, called Ether (ETH) or Ethereum.

• *Banking*: By nature, banking operations such as getting a loan or a mortgage, or even processing payments are slow and frustrating. Due to limitations of traditional banking like minimum balance requirements, limited access, and banking fees, many consumers are searching for banking alternatives. Blockchain may offer a hassle-free alternative to traditional banking. Banks and other financial institutions are already using Blockchain to optimize their services, cut back on fraud, and reduce fees for customers. Banks can leverage smart contracts to facilitate repetitive operations and payments, making the overall workflow go smoother and a way faster. Blockchain allows you to eliminate intermediaries in financial transactions, reduce costs and optimize many processes. For example, it allows banks to gain access to a common database of fraudsters and prevent money laundering. The efficiency of the banking system is also increased by reducing costs. BC provides fast and cheap money transfers. Many banks are using Blockchain trade finance platforms to create smart contracts between participants, increasing efficiency and transparency. Instead of trusting individual intermediaries like banks, participants can put their trust in the accuracy and security of the distributed ledger itself. The shift from interpersonal trust to system trust reduces uncertainty and counterparty risk. Some experts believe that Blockchain will eventually replace or complement traditional banking systems. Examples of US banks using Blockchain include Bank of America, JP Morgan Chase, Wells Fargo, and Goldman Sachs groups, as illustrated in Figure 3.3 [16].

Figure 3.3 Examples of US banks using Blockchain [16]

- *Accounting*: Reliable and up-to-date accounting records between counterparties make the audit process more transparent and significantly reduce time. The automation of processes did not lead to the disappearance of the auditors or accountants. But their role in the company will change. The growing popularity of Blockchain has led to services such as Blockchain consulting and Blockchain solution audits.

- *Smart Contracts*: Perhaps the most impactful application of Blockchain in finance is its ability to efficiently establish trust through smart contracts. The Blockchain technology provides an excellent foundation for smart contracts, which are programs stored on the Blockchain network that run when predetermined conditions are met. Smart contracts are similar to physical contracts, except the stipulations of the contract are fulfilled in real time via the Blockchain. Using smart contracts, Blockchain can automate and execute complex trade finance agreements, reducing the risk of fraud and error. Smart contracts enabled by Blockchain technology can help all parties create legally binding financial agreements that they will execute with a guarantee once all prerequisites have been satisfied. For example, in the law sector, Blockchain technology can be used to create smart contracts for legal processes. These smart contracts must be solidly grounded in the law and adhere to all applicable regulations. They generally have the potential to boost data trust.

• *Money Transfers*: From the beginning with Bitcoin, Blockchain technology was designed to move funds from point A to point B without a central governing body. As Blockchains have evolved, they have been able to achieve much faster and cheaper transactions. One prominent example is Ripple, a company that uses Blockchain technology for RippleNet, a global payments network. RippleNet transactions process within five seconds and cost just a fraction of a cent. Although some cash will always remain in circulation, the concept of "money" in the future will be very different from money as we conceive of it now.

• *Credit Reporting*: Credit reports dramatically impact customers' financial lives. Blockchain-based credit reporting is more secure than traditional server-based reporting. Blockchain may also enable companies to take non-traditional factors into account when calculating credit scores.

• *Digital Identity*: Financial institutions are responsible for maintaining the integrity of a customer's digital identity, comprising some of our most sensitive information. We trust banks with safeguarding our passport information, biometric scans, social security number, accounts and addresses. Current identity management solutions are no longer effective. Blockchain technology solves this problem by offering secure and decentralized digital identity management. Blockchain technology uses encryption and private keys to protect personal information, making hacking almost impossible.

• *Trade Finance*: This is a process whereby a financial institution grants a credit facility in order to ensure security over the transfer of goods. It is another Blockchain application in financial services. It is one of the oldest and largest markets in the world. But it is outdated, relying heavily on manual paperwork, emails, and phone calls. This opens the door for inefficiencies, fraud, and higher costs. Trading financing requires many international rules and regulations that regulate the activities of importers and exporters. It involves so many entities (exporter, importer, their respective banks, the shipping company, insurance companies, clearing and forwarding agents, etc.) and it is difficult for all of them to have a consistent view of the data. Trade finance can become essentially challenging when transactions in the form of assets have to be recorded with a clear date and time stamp. Blockchain has the potential to revolutionize trade finance by offering transparency, security, speed, and lower fees. A bank with a rich data set can turn this data into valuable information for its clients with the aid of Blockchain. Figure 3.4 compares traditional trade finance with Blockchain-based trade finance [17].

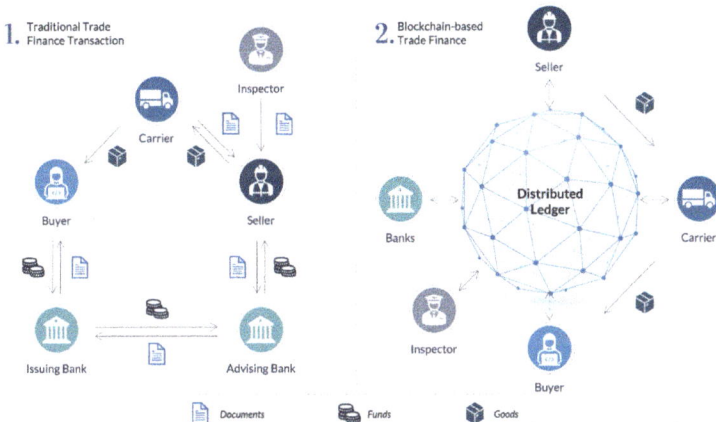

Figure 3.4 Comparing traditional trade finance with Blockchain-based trade finance [17].

• *Cross-border Transactions*: The traditional cross-border payment systems are fraught with high fees, long processing times, and intermediaries.

The entire area of cross-border payment with its legacy systems is an attractive target for Blockchain technologies. When used for cross-border transactions, Blockchain can make the process faster, more accurate, and less expensive. Trade finance helps facilitate international trade by providing funds for importers and exporters to engage in global trade transactions.

- *Insurance*: Blockchain technology's potential to revolutionize the insurance industry lies in its ability to streamline claims processing, enhance transparency, and reduce fraud. By storing policy and claims data on a Blockchain, insurers can automate the claims process using smart contracts, leading to faster payouts and reduced administrative costs. Operations can be performed faster and be less vulnerable to occasional mistakes

- *Credit Score*: Banks and other financial institutes require an applicant's credit score before proceeding with a loan application. One limitation of the current credit management system is that the current credit score of a person does not remain valid in a different country. Therefore, a universal credit score is needed. Managing credit score using Blockchain could bring transparency to the system. Blockchain allows lenders to access the immutable records of financial transactions to understand the creditworthiness of a person.

- *Stock Exchange*: Trading activities are dependent on trust. The current stock market involves entities like regulators, brokers, and the stock exchange that add more cost to the system. A decentralized approach to manage the stock exchange can make the system highly efficient. Blockchain can eliminate the need for third-party regulators as regulations can be built on smart contracts.

- *Asset Management*: The traditional approach to asset management is slow and complicated since it involves many intermediaries. Blockchain tackles this by enabling stakeholders to manage their digitized assets directly, with no external assistance. It is especially useful for international companies that have to comply with different jurisdictions. The asset management sector majorly focuses on a centralized digital system that provides a real-time visibility of assets within the systems.

Some of these services or applications of Blockchain in finance industry are shown in Figure 3.5 [18].

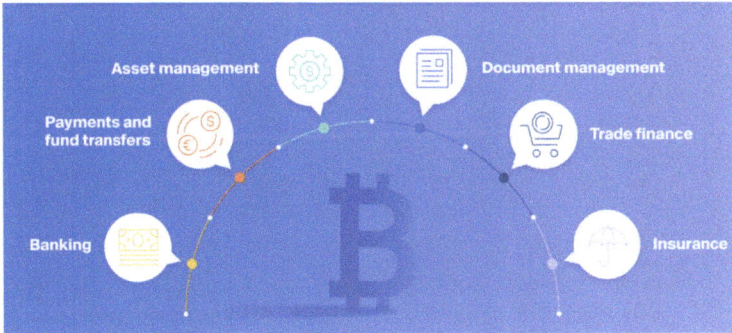

Figure 3.5 Some applications of Blockchain in finance [15].

3.4 BENEFITS

In the financial services industry, Blockchain helps to completely revise the existing structure of banks, speed up transactions, and modernize stock exchanges with proper security. Blockchain technology is known to improve payment transparency, efficiency, trust, and security. It is also known to reduce the cost for financial service firms and their users. Blockchain technology is important for companies since its adoption guarantees revenue growth, lower costs, and increased efficiency. It promises to provide greater efficiency, lower transaction costs, better transparency, a speedier rate of financial innovation, and a much lower carbon footprint. It has the potential to improve client affordability, reduce fraud risk, and increase transparency in the financial services sector. Blockchain may ultimately disrupt paper-clogged industries, such as health care and insurance. The BC technology has the following additional benefits [19]:

- *Security*: The first priority for any financial body is in the area of security. Given the tight and popular security pattern Blockchain holds, most banks are obviously going to use this in storing assets that are of extreme value. Its distributed consensus based architecture eliminates single points of failure and reduces the need for data intermediaries such as transfer agents, messaging system operators and inefficient monopolistic utilities.

- *Immutability*: A key benefit of Blockchain technology is that it is immutable, meaning that once something has been written to the Blockchain it cannot be altered. This is a key advantage in the financial services industry where data integrity is crucial. Before the invention of bitcoins, anything digital could be copied.

- *Trust*: Industries that can benefit the most from Blockchain are those where trust among participants is low, and the need for record security and integrity is high. Blockchain's transparent and immutable ledger makes it easy for different parties in a business network to collaborate, manage data, reach agreements, and creating increased trust and efficiency.

- *Privacy*: It provides market-leading tools for granular data privacy across every layer of the software stack, allowing selective sharing of data in business networks. This dramatically improves transparency, trust and efficiency while maintaining privacy and confidentiality.

- *High-Performance*: It's private and hybrid networks are engineered to sustain hundreds of transactions per second and periodic surges in network activity.

- *Boost Productivity*: Blockchain technology has the potential to revolutionize the financial industry by boosting productivity, transparency, and security, cutting costs, and spurring a previously unheard-of wave of innovation.

- *Scalability*: It supports interoperability between private and public chains, offering each enterprise solution the global reach and tremendous resilience.

- *Economic Benefits*: Automated, more efficient processes trigger reduced infrastructure costs, operation costs, and transaction costs. From a business perspective, Blockchain technology largely centers on cost savings and new revenue from services parameters across all industries.

- *Cost Reduction*: Blockchain enables cost reduction and time efficiency in many ways for trade finance processes. Eliminating intermediaries means lower transaction fees and commissions for traders. Automating tasks through smart contracts reduces the need for manual labor, reviews, and approval.

- *Payment*: Given that Blockchain can be used in both domestic and international payment, most banks have started keying into the idea of using Blockchain for payment.

- *Mobile Money*: Services involving the exchange of money are now being performed on a Blockchain platform. This is disrupting related services provided today in two major industries, namely Financial Services and Telecommunications, as both provide money transfer, remittances and bill payment services. Blockchain is already being used to provide payment remittance and transfer services, although it is not yet widely.

- *Reduced Fraud*: Involving money in any situation leads to increased chances of fraudulent activities. Much of the fraud connects with the enormous complexity and inefficiency of information exchange between market participants. Using Blockchain technology, all of the data associated with a transaction is recorded, simple to obtain and non-modifiable. This means it will greatly reduce the risk of incorrect classification, error, and financial statement fraud.

Some of these benefits are shown in Figure 3.6 [20].

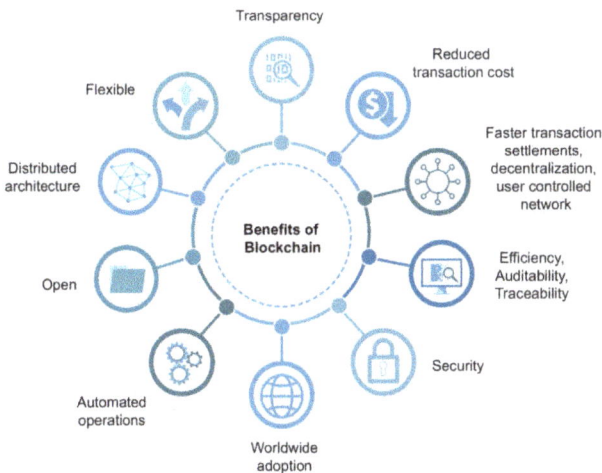

Figure 3.6 Some benefits of Blockchain [20].

3.5 CHALLENGES

While Blockchain offers many benefits, there are still challenges and limitations that may slow its adoption for trade finance. The challenges include the following [21,22]:

- *Immaturity*: Blockchain is still an emerging technology. Issues around scalability, security, and integration need to be improved before wide adoption.

- *Regulatory Uncertainty*: The financial industry is subject to several complex regulations, making regulatory compliance a significant challenge. Regulators are not often clear in their regulatory stance on the new technology, and obtaining their clearance is not always easy. Unclear regulations around Blockchain could slow its adoption. Regulators need to develop frameworks to govern Blockchain systems. The use of Blockchain technology may be subject to regulatory scrutiny. Regulators need to be convinced that the use of Blockchain technology is safe, secure, and compliant with applicable regulations.

- *Standardization*: To implement Blockchain technology effectively, there needs to be standardization across all parties, including the development of common protocols, data formats, and interfaces.

- *High Costs*: Implementing Blockchain networks require high initial investments. The costs of integrating with legacy systems can be significant.

- *Complexity*: Blockchain systems involve complicated technical specifications that require expertise to develop and maintain. This complexity limits the number of players.

- *Single Point of Failure*: The entire Blockchain network is reliant on internet connectivity. An outage can disrupt the whole system.

- *Compatibility Issues*: Blockchain networks developed by different players may not be compatible, limiting their usefulness.

- *Energy Consumption*: The computational power required

to run Blockchain networks consumes large amounts of energy. This may not be sustainable.

• *Participation*: Widespread adoption of Blockchain requires most participants to join the network. Getting competing parties to collaborate can be challenging.

• *Data Privacy*: People trust banks and financial institutions for storing their funds. In order for Blockchain to take their place, it is important to ensure that the data stored on the Blockchain technology is kept securely and would not hamper the identity of any individual. Public Blockchains expose all transaction details, raising privacy concerns. Solutions to address this are still limited.

• *Interoperability*: Interoperability between Blockchain systems is a challenge. The Blockchain technology is not bounded by any international rules and regulations that place a standard to it. To achieve the benefits of Blockchain, different Blockchain networks need to be able to communicate with each other. However, there is currently no widely accepted standard for interoperability, and different Blockchain platforms may have different protocols, making it challenging to integrate them. Blockchain technology has not yet attained the highest level of interoperability in the financial sector due to energy consumption, privacy ethics, user trust, laws, regulations, compliance rules/protocols, supervision, and network integration.

• *Scalability*: Blockchain technology is still relatively new, and it can be challenging to scale it up to handle large volumes of transactions. This can be particularly challenging in trade finance, where there are often high volumes of transactions that need to be processed quickly.

Some of these key challenges are illustrated in Figure 3.7 [22]. For the Blockchain to succeed in mainstream finance, these challenges must be addressed and overcome.

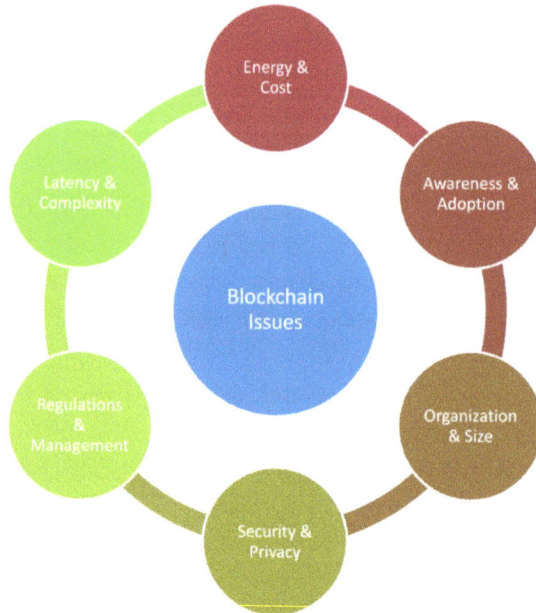

Figure 3.7   Key challenges for Blockchain adoption [22].

3.6 CONCLUSION

Blockchain, a decentralized digital ledger, records transactions across computer networks. It is a tamper-proof log of sensitive activities that are efficiently and securely created. It has immense potential to transform the finance sector. It will not take long before Blockchain-based finance systems becomes the norm rather than the exception.

Blockchain technology is known to improve payment transparency, efficiency, trust, and security. It is also known to reduce the cost for financial service firms and their users. Blockchain technology has disrupted the finance industry by introducing decentralized, secure, and transparent ways of conducting financial transactions and managing assets. As the technology continues to mature, it is likely to have even more profound effects on how financial services are structured and delivered.

Financial service providers find Blockchain technology useful to enhance authenticity, security, and risk management. They are generally optimistic about using Blockchain in the financial sector. Decentralized finance was made possible by the use of Blockchain in financial services. Blockchain is being recognized as the new technology that would reduce fraud in the financial world.

Blockchain is still an evolving and therefore immature technology. In spite of this, more and more industry giants are investing in Blockchain technology. The future of Blockchain technology in the finance industry looks promising as more and more financial institutions are using Blockchain to streamline their operations, reduce costs, and improve security. More information about Blockchain in finance can be found in the books in [23-34] and related journal: International Journal of Finance, Insurance and Risk Management.

REFERENCES

[1] M. Javaid et al., "A review of Blockchain technology applications for financial services," Bench Council Transactions on Benchmarks, Standards and Evaluations, vol. 2, no. 3, July 2022.

[2] "Why Blockchain is the future of finance and how it affects the global economy,"

https://medium.com/unidocore/why-Blockchain-is-the-future-of-finance-and-how-it-affects-the-global-economy-8e24c33ade0b

[3] C. Baker, "Blockchain in financial services," 2020,

https://papers.ssrn.com/sol3/papers.cfm?abstract_id=3578599

[4] M. N. O. Sadiku, U. C. Chukwu, and J. O. Sadiku, "Blockchain in finance," International Journal on Economics, Finance and Sustainable Development, vol. 5, no.9, September 2023, pp. 20-30.

[5] M. N. O. Sadiku, Y. Wang, S. Cui, and S. M. Musa, "A primer on Blockchain," International Journal of Advances in Scientific Research and Engineering, vol. 4, no. 2, February 2018, pp. 40-44.

[6] A. Mia et al., "Blockchain in financial services: Current status, adaptation challenges, and future vision," https://www.worldscientific.com/doi/10.1142/S0219877023300045

[7] S. Depolo, "Why you should care about Blockchains: the non-financial uses of Blockchain technology," March 2016,

https://www.nesta.org.uk/blog/why-you-should-care-about-Blockchains-non-financial-uses-Blockchain-technology

[8] O. Bheda, "What is Blockchain?" https://builtin.com/Blockchain

[9] M. Iansiti and K. R. Lakhani, "The truth about Blockchain," Harvard Business Review, Jan./Feb. 2017.

https://hbr.org/2017/01/the-truth-about-Blockchain

[10] W. T. Tsai et al., "A system view of financial Blockchains," Proceedings of IEEE Symposium on Service-Oriented System Engineering, 2016, pp. 450-457.

[11] "5 Common Blockchain applications in financial services,"

https://www.hydrogenplatform.com/blog/5-common-Blockchain-applications-in-financial-services

[12] "Blockchain in finance: What it is and how it's used,"
https://builtin.com/Blockchain/Blockchain-banking-finance-fintech

[13] L. Daly, "Uses for Blockchain in the financial services industry," July 2022,
https://www.fool.com/investing/stock-market/market-sectors/financials/Blockchain-stocks/Blockchain-in-finance/

[14] "The power of Blockchain technology and its revolutionary uses in the financial sector," February 2022,
https://www.salesforce.com/eu/blog/2020/02/how-financial-services-are-implementing-Blockchain-technology.html

[15] "How can the Blockchain be used for financial services?"
https://upplabs.com/blog/how-can-the-Blockchain-be-used-for-financial-services/

[16] M. Pratap, " How is Blockchain revolutionizing banking and financial markets," July 2018,
https://hackernoon.com/how-is-Blockchain-revolutionizing-banking-and-financial-markets-9241df07c18b

[17] "Revolutionizing trade finance with Blockchain technology," February 2020,
https://mahanakornpartners.com/revolutionizing-trade-finance-with-Blockchain-technology/

[18] "The future of financial services: 6 impactful Blockchain use cases in finance," September 2022,
https://pixelplex.io/blog/Blockchain-in-finance/

[19] "What are the benefits of Blockchain in finance?"
https://consensys.net/Blockchain-use-cases/finance/

[20] "Knowledge byte: The real benefits of Blockchain,"
https://www.cloudcredential.org/blog/who-participates-in-a-Blockchain-network-and-what-are-its-benefits/

[21] "Exploring the benefits of Blockchain in trade finance,"
https://vegavid.com/blog/Blockchain-in-trade-finance-benefits/

[22] "Blockchain technology: Challenges and limitations," https://www.researchgate.net/figure/Blockchain-technology-challenges-and-limitations_fig2_322814269

[23] H. Mohamed and H. Ali, Blockchain, Fintech, and Islamic Finance: Building the Future in the New Islamic Digital Economy. De Gruyter, 2018.

[24] S. S. Smith, Blockchain, Artificial Intelligence and Financial Services: Implications and Applications for Finance and Accounting Professionals. Springer, 2020.

[25] E. Hofmann, U. M. Strewe, and N. Bosia, Supply Chain Finance and Blockchain Technology: The Case of Reverse Securitisation. Springer, 2017.

[26] D. L. K. Chuen and R. H. Deng. Handbook of Blockchain, Digital Finance, And Inclusion: Cryptocurrency, Fintech, Insurtech, Regulation, Chinatech, Mobile Security, And Distributed Ledger. Academic Press, 2017.

[27] A. Tapscott, Financial Services Revolution: How Blockchain is Transforming Money, Markets, and Banking. Barlow Book Publishing, 2020.

[28] P. Martino, Blockchain and Banking: How Technological Innovations Are Shaping the Banking Industry. Springer, 2021.

[29] I. Roy, Blockchain Development for Finance Projects: Building Next-generation Financial Applications Using Ethereum, Hyperledger Fabric, and Stellar. Packt Publishing, 2020.

[30] U. Hacioglu (ed.), Blockchain Economics and Financial Market Innovation

Financial Innovations in the Digital Age. Springer, 2019.

[31] M. C. M. Metzger, Blockchain Banking: The Future Of Money and Finance. Books on Demand, 2023.

[32] R. Hayen, Blockchain & FinTech: A Comprehensive Blueprint to Understanding Blockchain & Financial Technology. CreateSpace Independent Publishing Platform, 2017.

[33] P. Gaffney, K. Sonlin, and H. Konings, Your Ultimate Guide to the Tokenization of Finance. Authors Unite Publishing, 2023.

[34] S. Khan et al., Blockchain Technology and Computational Excellence for Society 5.0. IGI Global, 2022.

CHAPTER 4

BLOCKCHAIN IN HEALTHCARE

"The whole point of using a blockchain is to let people — in particular, people who don't trust one another — share valuable data in a secure, tamperproof way."

— MIT Technology Review

4.1 INTRODUCTION

Health is the foundation of a happy life. Healthcare is an important part of life. The healthcare industry is one of the world's largest industries and is resistant to change and innovative practices. Modern healthcare systems have become highly complex and costly. The cost of healthcare delivery is continuously rising, causing a crisis. The healthcare industry is under extreme pressure to both curb the rising cost of healthcare and provide high quality to patients. Currently, sensitive medical records lack a secure structure, causing data breaches. Blockchain technology has some interesting properties, such as its decentralized nature, immutability, decentralization, transparency, and permissionless, that may provide the solution by addressing pressing issues in healthcare [1,2]. Figure 4.1 shows how Blockchain meets healthcare requirements [3].

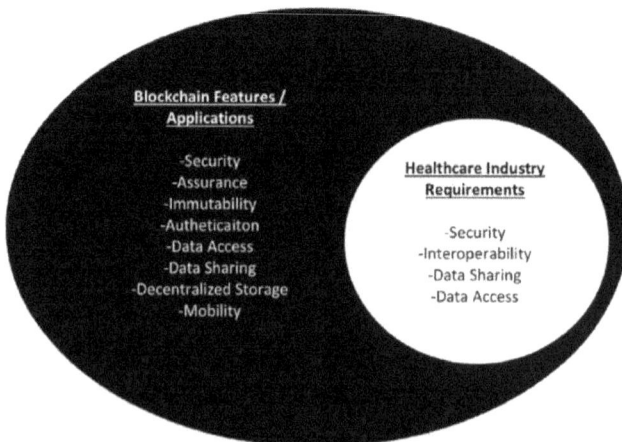

Figure 4.1   How Blockchain meets healthcare requirements [3].

The Blockchain technology is disruptive and considered as the fourth industrial revolution that will change the world.

Blockchain (BC) consists of a shared or distributed database used to maintain a growing list of transactions, called blocks. Blockchain technology, often called the chain of trust, can support transactional applications and streamline business processes by establishing trust, accountability, and transparency. The so-called digital ledger technology was developed in 2008 by Satoshi Nakamoto, who designed it as the underpinning for the exchange of the digital cryptocurrency known as Bitcoin. BC forms the backbone of cryptocurrencies like bitcoin, Litecoin, and Ethereum. They work by keeping track of transactions in a distributed ledger.

Although Blockchain was first largely applied in financial industry as the technology that allowed Bitcoin to operate, it has applications for many industries including healthcare, insurance, pharmacy, manufacturing, healthcare, e-voting, legal contracts, tourism, energy, and travel industry. Healthcare will benefit from the early work in finance and leverage Blockchain applications in finance. Applying BC in healthcare serves to improve patient care. BC technology offers patients and care-givers the ability to securely share patient identity and healthcare information across platforms. Imagine a future where patients hold the keys to their healthcare passport. Imagine a better quality of care for both patients and care providers [4].

As a catalyst for change, the Blockchain technology is going to change healthcare in major ways. The decentralized ledger can be used to store personal details of the patient. The main motivation for using Blockchain in healthcare is to solve the data integrality, data interoperability, and privacy issues in current health systems. Blockchain eliminates the need of a middleman who plays the role of verifying transaction in the healthcare industry [5].

This chapter provides an introduction to the use of Blockchain in healthcare. It begins by explaining how Blockchain works. It covers the two types of Blockchain. It discusses some of its applications in healthcare industry. It covers global Blockchain healthcare. It addresses the benefits and challenges of Blockchain in healthcare. The last section concludes with comments.

4.2 OVERVIEW OF BLOCKCHAIN

Blockchain technology is a permanent record of online transactions. It is a distributed tamper-proof database, shared, and maintained by multiple parties. It is a new enabling technology that is expected to revolutionize many industries, including healthcare. It has the potential for addressing significant healthcare issues. The BC technology allows participants to move data in real-time, without exposing the channels to theft, forgery, and malice.

The term "Blockchain" refers to the way BC stores transaction data – in "blocks" that are linked together to form a "chain." The chain grows as the number of transactions increases. Since every entry is stored as a block on a chain, the care you receive is added to your personal ledger. The first Blockchain was conceived in 2008 by an anonymous person or group known as Satoshi Nakamoto, who published a white paper introducing the concept of a peer-to-peer electronic cash system he called Bitcoin [6].

At its core, Blockchain is a distributed system recording and storing transaction records. In a Blockchain system, there is no central authority. Instead, transaction records are stored and distributed across all network participants. Rather than having a centrally located database that manages records, the database is distributed to the networks and transactions are kept secure via cryptography. BC eliminates the need for a middleman that traditionally may facilitate such transactions.

Fundamentally, Blockchains are distributed digital database that record and maintain a list of transactions taking place in real time. They may also be regarded as decentralized ledgers that sequentially record transactions or interactions among users within a distributed network. They have the following properties [7]:

- Firstly, they are autonomous. They run on their own, without any person or company in charge.
- Secondly, they are permanent. They are like global computers with 100 percent uptime. Because the contents of the database is copied across thousands of computers, if 99 per cent of the computers running it were taken offline, the records would remain accessible and the network could

rebuild itself.

• Thirdly, they are secure and tamper-proof. Each record in Blockchain is time stamped and stored cryptographically. The encryption used on Blockchains like Bitcoin and Ethereum is industry standard, open source, and has never been broken.

• Fourthly, they are open, allowing anyone to develop products and services on them.

• Fifthly, as Blockchain is a shared system, costs are also shared between all of its users.

The Blockchain was designed so transactions are immutable, i.e. they cannot be deleted. Thus, Blockchains are secure and meddle-free by design. Data can be distributed, but not copied. When it comes to digital assets and transactions, you can put almost anything on a Blockchain. Different scenarios call for different Blockchains. Figure 4.2 depicts some general applications of Blockchain [8].

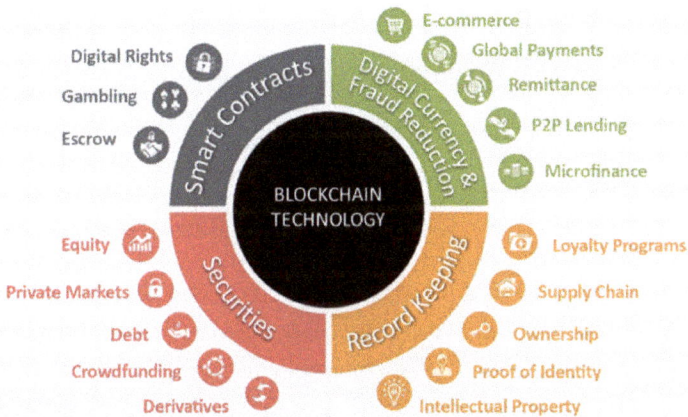

Figure 4.2   General applications of Blockchain [8].

The BC technology currently has the following features [9,10]:

1. *Peer-to-Peer (P2P) Network*: The first requirement of BC is a network, an infrastructure shared by multiple parties. This can be a LAN at a small scale or the Internet at a large scale. All nodes participating in a BC are connected in a decentralized P2P network. Transactions are broadcast to the P2P network. Due to some limitations of P2P networks, some vendors have provided cloud-based BCs.

2. *Cascaded Encryption*: A BC uses encryption to protect transaction data. Blocks are encrypted in a cascaded manner, i.e. the encryption result of the previous block is used in encrypting the current block. The BC is secured by public key cryptography, with each peer generating its own public-private key pairs.

3. *Distributed Database*: A BC is digitally distributed across a number of computers. Each party on a BC has access to the entire database and no single party controls the data or the information. Since BC is decentralized, there is no need for central authorizes such as banks.

4. *Transparency with Pseudonymity*: Each node or participant on a Blockchain has a unique 30-plus-character alphanumeric address that identifies it. Users can choose to remain anonymous or provide proof of their identity to others.

5. *Irreversibility of Records*: Once a transaction is entered in the database and the accounts are updated, the records cannot be altered. Records on the database is permanent, chronologically ordered, and available to all others on the network.

4.3 TYPES OF BLOCKCHAINS

There are two types of Blochains: public and private. Public Blockchains are cryptocurrencies such as Bitcoin, enabling peer-to-peer transactions. Private Blockchains use Blockchain-based platforms such as Ethereum or Blockchain-as-a-service (BaaS) platforms running on private cloud infrastructure. A private BC is an intranet, while a public BC is the Internet. Companies will be disrupted the most by public Blockchains.

BCs may be permissioned or permissionless. In a permissioned BC, each participant has a unique identity. Permissionless BCs are appealing because they allow anyone to join, participate or leave the protocol execution without seeking permission from a centralized or distributed authority. However, permissionless BCs, such as Ethereum or Bitcoin, face transaction volume constraints. Both permisisoned and permisionless can be implemented in healthcare [11].

Blockchain technology is emerging and evolving. Its three

generations are mentioned here. It was first proposed to support cryptocurrencies like Bitcoin, so cryptocurrency Blockchains are known as Blockchain 1.0. All applications of Blockchain in the financial area made possible by the union of ethereum smart contracts with digital currencies are labelled Blockchain 2.0. This generation allows customers have a transaction of stock, bill of exchange, intellectual property right or anything related to smart contract. All applications of Blockchain technology referable to the wider spectrum of non-cryptocurrency-related uses are usually known as Blockchain 3.0 applications. Blockchain 3.0 applications include electronic voting, healthcare, education, identity management, and decentralized notary [12,13].

MedRec has been proposed as a novel, decentralized record management system to handle electronic health record (EHRs) using Blockchain technology. It manages authentication, confidentiality, accountability, and data sharing. It enables patient data sharing and incentives for medical researchers to sustain the system [14] Healthchain is developed on the foundation of Blockchain using IBM Blockchain initiative.

4.4 APPLICATIONS

A typical Blockchain healthcare system is shown in Figure 4.3 [15].

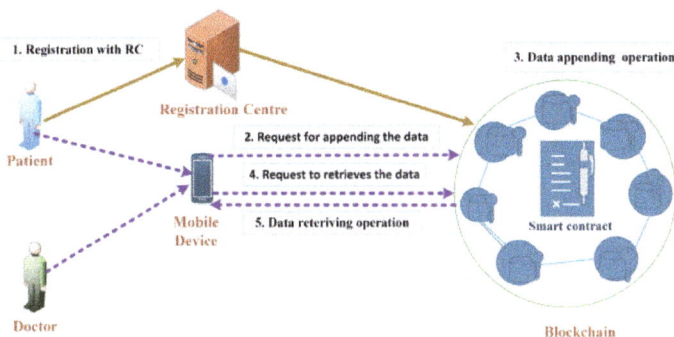

Figure 4.3 A typical Blockchain healthcare system [15].

Blockchain has a wide range of applications in healthcare. It has the potential for addressing significant healthcare issues. Several areas of healthcare can be enhanced using Blockchain technologies including medical records or HER systems, device tracking, clinical trials, pharmaceutical tracing, and health insurance. The following are the most likely applications [16]:

- *Medical Data Management*: Healthcare is a data-intensive industry. The healthcare industry is drowning in data— patient medical records, complex billing, clinical trials, medical research, etc. The goal of BC is to give patients and their providers one-stop access to their entire medical history across all providers. Blockchain is able to securely, privately and comprehensively track patient health records. It makes ERHs more efficient, disintermediated, and secure. Right now, a patient's medical records are dispersed across multiple providers and organizations due to the fact the health systems are fragmented into hospitals, community clinics, general practitioners, specialists, insurance departments, etc. Patient data is one valuable asset of patients, but patients have no control of their personal data. Some of the record pieces are with the primary doctor, some with specialists, and some on devices that track one's health. Every hospital and every doctor's office has a different way of storing the electronic medical records (EMRs), also known as electronic health records (EHRs). For example, in the city of Boston alone, there are 26 different EMRs, each with its own language for representing and sharing data. This situation is costing us money, professional burnout, and sometimes even lives. Blockchain can help us assemble all of these pieces in real-time. This way care providers can have the complete medical history of the patient. For healthcare to reap the benefits of a Blockchain-based medical record, it must grant access to everyone that might need patient's information [17,18]. Blockchain can also create a mechanism to access EHRs stored on the cloud.

- *Drug Development*: Blockchains can facilitate new drug development by making patient results more widely accessible. It can help reduce the counterfeit drug implications. The issue of counterfeit medicines has become increasingly pressing in view of the economic cost of the global black market and the risk to human life that comes from taking counterfeit drugs. Blockchain technology is an excellent counter to threats that are rapidly approaching (integrity-based attacks) and it is a good forward-looking tool we might deploy to address them. BC will also enable drug developers to run clinical trials and share medical samples more securely [19].

• *Clinical Trials*: Using Blockchain can make clinical trials reliable at each step by keeping track and time-stamping at each phase of the trial. This could reduce waste. Another Blockchain use-case would be the adoption of electronic informed consent in clinical trials. BC improves accountability and transparency in the clinical trial reporting process.

• *Data Security*: The perpetrators can steal credit card, banking information, and health records. Sensitive data must be kept safe from hackers and intruders. Blockchain technology has the potential to be the infrastructure that is needed to keep health data private and secure. It makes health information exchanges (HIE) more secure, efficient, and interoperable. BC requires no one central administrator, and it has unprecedented security benefits because records are distributed across a network that are always in sync. For example, Factom employs Blockchain technology to securely store digital health records.

• *Pharmaceutical Sector*: Pharmaceutical companies need to have a very secure supply chain because of the kind of product they deal with. Blockchain in healthcare helps pharmaceutical companies create auditable, unalterable, secure, and distributed databases for storing and accessing drug trial data. Blockchain has serious implications for pharmaceutical supply chain management. By scanning the supply chain, the company's app lets patients know if they are taking falsified medicines. Through its app, the company's Blockchain-based system can help prevent patients from taking counterfeit medicines. Blockchain can help overcome the increasing risks around counterfeit and unapproved drugs.

• *Medication Adherence*: This is crucial for patient safety. There are Blockchain firms like Guardtime, a UK based health tech company, that are already using BC technology to ensure medication adherence.

- *Fraud Detection*: There are risks of using the digital system: fraud and forgery. Fraud detection refers to the process of verifying a document to identify any tampering with the information. Blockchain technology can be in the field of pharmacy with the purpose of detection of counterfeit and poor-quality medicinal products [20].

Other applications of Blockchain healthcare include counterfeit drug prevention, validation and payment of claims, clinical trial results, outcome-based payments, reimbursement of healthcare services, exchange of health data, and supply chains [21].

4.6 BENEFITS

The main benefits of Blockchain in healthcare are data interoperability, security, efficiency, and accessibility. Ownership and privacy of data are important issues that Blockchain could solve. It holds the promise to unite the disparate healthcare processes, reduce costs, improve regulatory compliance, improve patient experience, provide healthcare at lower costs, and autonomous monitoring and preventive maintenance of medical devices. It ensures the reliability of the stored data, which is determined by the fact that each record is confirmed from several sources. It provides a shared and transparent history of all the transparency, and immutability. Some of these benefits are illustrated in Figure 4.4 and explained as follows [22].

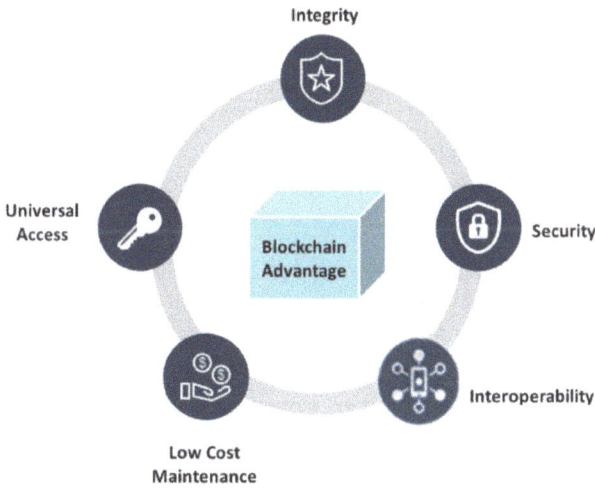

Figure 4.4   Some of the benefits of healthcare Blockchain [22].

• *Security*: One of the greatest benefits of Blockchain technology is that it is significantly more secure than other data storage platforms. Healthcare generates a massive amount of confidential data that can be detrimental if it falls in the wrong hands. Security should include protecting the confidentiality, integrity, and availability of sensitive data and systems. Blockchain can improve data security since it has no centralized point of failure. This makes it appealing to store and share personal health information (PHI) and stop Cyber-attacks. In January of 2017, IBM Watson partnered with the FDA to develop a secure exchange of health data utilizing Blockchain technology. Although the technology provides resilience to certain types of attacks but it is by no means entirely secure. Cybercriminals deliberately target the healthcare industry for the valuable information that they store for individuals. This valuable information includes the names, birth dates, and social security numbers.

• *Interoperability*: This refers to automatic and seamless exchange of health information across health information systems. Seamless exchange of health data across healthcare systems would be advantageous. In healthcare, interoperability allows two or more systems to exchange and use information. BC can allow improved interoperability as data across multiple systems can be exchanged and accessed simultaneously. It can enhance interoperability across a global market, eliminating system boundaries and geographic limitations. Improved interoperability or cost-effectiveness can result in efficiency [23].

• *Integrity*: BC can help bring some of the data integrity, accessibility, security, privacy, and interoperability needed by pharmaceutical companies, and eliminate falsified medication. Blockchain technology helps ensure data integrity while encryption of data enhances data security across the network. The immutability of a Blockchain carries with it inherent integrity, as blocks cannot be rewritten without collaboration of a majority of nodes.

• *Universal Access*: Blockchain provides a possible future solution for data sharing. It has the potential to reduce healthcare costs, streamline business processes, and improve access to information. It ensures that required data is present at every node and is available for use to the authorized entities. Data sharing enables real-time updates. It also enables collaborative clinical decision-making in telemedicine and precision medicine. Healthcare organizations need not compete among themselves because they all have access to the same information. However, the universal availability of patient data poses some challenges [24].

• *Patient-Centered Care*: The healthcare industry is shifting from volume-based care to value-based care that promotes patient-centered care with higher quality. In patient-centered care, patients are given access to their clinical data in real-time, with a comprehensive view of their entire health history. There are ongoing efforts to explore the use of Blockchain-based systems to create a more patient-centric environment wherein patients would be in control of their own health data. Interoperability is crucial to support a patient-centric model [25].

• *Cost Saving*: Blockchain technology saves the cost of mediation, as a Blockchain involves no mediator. The interfacing of different systems would also save costs.

Blockchain will speed up the R&D cycle and time to market of new drugs. Blockchain technology has the potential to transform healthcare systems because it places the patient at the center of the health care ecosystem. BC is the perfect solution when we need to document a patient's health record or to secure the movement of drugs through the supply chain. BC has the potential of transmitting patent record across geographies without compromising its integrity, privacy, and security. Because of these benefits, Blockchain is beginning to declare itself as a potentially game-changing technology in healthcare.

4.7 CHALLENGES

Although Blockchain presents many opportunities for healthcare, it is not fully mature yet. Several technical challenges must be addressed before a healthcare Blockchain can be adopted nationwide and worldwide[3,26].

- *Integration*: Integrating Blockchain into healthcare is an uphill task. The success of in the healthcare industry depends on whether hospitals, clinics, and other organizations are willing to cooperate in building the technical infrastructure required. The system must facilitate the exchange of sensitive health information between patients and providers as well as exchanges between providers, while remaining secure from malicious attacks [27].

- *Data Ownership and Privacy*: Transferring data ownership from the government and organizations to patients make patients to become active agents in their own care, but it would require extensive transformation of legacy systems. Enabling direct patient involvement in controlling the secure use of their records will ensure patient privacy and potentially lead to improved health outcomes. Blockchains are not ideal for storing private information due to the transparency that they provide. Data privacy and the ability to access sensitive patient information are the key challenges in the design of a healthcare Blockchain application [28]. As they work today, anyone can look at the bitcoin or Ethereum ledger at any time. If someone can identify your records on the Blockchain, they know everything about your medical history. By design, BC technology is distributed and storage space is limited, so small data or metadata is preferable.

- *Regulation*: Some healthcare professionals think that Blockchain deployments are held back by regulatory issues.

- *Lack of standardization*: As a relatively new technology, there is a lack of standardization. This hinders its broad acceptance and slows down development.

Critics question the scalability, security, and sustainability of Blockchain technology. The security of medical devices cannot be compromised in critical care. These challenges restricting the implementation of Blockchain technology need to be addressed.

4.8 CONCLUSION

Although the application of Blockchain to healthcare is in its infancy, its adoption has been exponential. The Blockchain revolution has made its way to the healthcare industry, and leaders are now wondering what is possible and how Blockchain can solve many issues that plague the industry. BC is the technology that will possibly have the greatest impact on the next few decades; not social media or big data or robotics. Although BC is not fully mature, the healthcare system can take advantage of a beneficial disruptive innovation that will stand the test of time like Blockchain.

The rapid growth in the deployment of Blockchain healthcare warrants keeping a close eye on Blockchain in healthcare and the opportunities it will bring. Blockchain has great potential for the future and will cause disruptive changes in the healthcare industry [29]. For more information about Blockchain healthcare, one should consult the books in [30-42] and the following related journal: Blockchain in Healthcare Today .

REFERENCES

[1] H. S. Chen et al., "Blockchain in healthcare: A patient-centered model," Biomedical: Journal of Scientific & Technical Research, vol. 20, no. 3, 2019, pp. 15017-15022.

[2] M. Prokofieva and S. J. Miah, "Blockchain in healthcare," Australasian Journal of Information Systems, vol 23, 2019.

[3] T. McGhin et al., "Blockchain in healthcare applications: Research challenges and opportunities," Journal of Network and Computer Applications, vol. 135, 2019, pp. 62-75.

[4] S. Manski, "Building the Blockchain world: Technological commonwealth or just more of the same?" Strategic Change, vol. 26, no. 5, 2017, pp. 511-522.

[5] M. N. O. Sadiku, K. G. Eze, and S.M. Musa, "Block chain technology in healthcare," International Journal of Advances in Scientific Research and Engineering, vol. 4, no. 5, 2018, pp. 154-159.

[6] M. N. O. Sadiku, Y. Wang, S. Cui, and S. M. Musa, "A primer on Blockchain," International Journal of Advances in Scientific Research and Engineering, vol. 4, no. 2, February 2018, pp. 40-44.

[7] S. Depolo, "Why you should care about Blockchains: the non-financial uses of Blockchain technology," March 2016,

https://www.nesta.org.uk/blog/why-you-should-care-about-Blockchains-non-financial-uses-Blockchain-technology

[8] "Security in Blockchain applications,"

https://cri-lab.net/security-in-Blockchain-applications/

[9] M. Iansiti and K. R. Lakhani, "The truth about Blockchain," Harvard Business Review, Jan./Feb. 2017.

https://hbr.org/2017/01/the-truth-about-Blockchain

[10] W. T. Tsai et al., "A system view of financial Blockchains," Proceedings of IEEE Symposium on Service-Oriented System Engineering, 2016, pp. 450-457.

[11] Z. Alhadhrami et al., "Introducing Blockchains for healthcare," Proceedings of International Conference on Electrical and Computing Technologies and Applications, 2017.

[12] D. D. F. Maesa and P. Mori, "Blockchain 3.0 applications survey," Journal of Parallel and Distributed Computing, vol. 138, 2020, pp. 99-14.

[13] T. L. Nguyen, "Blockchain in healthcare: A new technology benefit for both patients and doctors," Proceedings of PICMET '18: Technology Management for Interconnected World, 2018.

 [14] A. Azaria et al., "MedRec: Using Blockchain for medical data access and permission management," Proceedings of the 2nd International Conference on Open and Big Data, 2016, pp. 25-30.

[15] V. Ramani et al., "Secure and efficient data accessibility in Blockchain based healthcare systems," Proceedings of IEEE Global Communications Conference, (GLOBECOM), December 2018.

[16] B. Marr, "This is why Blockchains will transform healthcare,"

https://www.forbes.com/sites/bernardmarr/2017/11/29/this-is-why-Blockchains-will-transform-healthcare/#467c7fbe1ebe

[17] "Blockchain in health care: The good, the bad and the ugly,"

https://www.forbes.com/sites/forbestechcouncil/2018/04/13/Blockchain-in-health-care-the-good-the-bad-and-the-ugly/#c00ec6462787

[18] D. V. Dimitrov, "Blockchain applications for healthcare data management," Healthcare Informatics Research, vol 25, no. 1, January 2019, pp. 51-56.

[19] "What the hell is Blockchain and what does it mean for healthcare and pharma?"

http://medicalfuturist.com/what-the-hell-is-Blockchain-what-does-it-mean-for-healthcare-and-pharma/

[20] V. Pashkov and O. Soloviov,"Legal implementation of Blockchain technology in pharmacy," SHS Web of Conferences, vol. 68, 2019.

[21] S. Attili, "Today's healthcare challenges and how Blockchain can help solve them – Join IBM for a HIMSS17 breakfast briefing,"

https://www.ibm.com/blogs/insights-on-business/healthcare/todays-healthcare-challenges-Blockchain-can-help-solve-join-ibmhimss17-breakfast-briefing/

[22] V. Rawal et al., "Blockchain: An opportunity to address many complex challenges in Healthcare," May 2018,

https://www.healthcare.digital/single-post/2018/05/26/Blockchain-An-opportunity-to-address-many-complex-challenges-in-Healthcare

[23] A. A. Vazirani et al., "Implementing Blockchains for efficient health care: Systematic review," Journal of Medical Internet Research, vol. 21, no. 2, February 2019.

[24] A. H. Mayer, C. A. da Costa, and R. R. Righi, "Electronic health records in a Blockchain: A systematic review ," Health Informatics Journal, 2019, pp. 1–16.

[25] P. Zhang et al., "Chapter one - Blockchain technology use cases in healthcare," Advances in Computers, vol. 111, 2018, pp. 1-41.

[26] M. A. Engelhardt, "Hitching healthcare to the chain: An introduction to Blockchain technology in the healthcare sector," Technology Innovation Management Review, vol. 7, no. 10, 2017, pp. 22-34.

[27] "Who will build the health-care Blockchain?"

https://www.technologyreview.com/s/608821/who-will-build-the-health-care-Blockchain/

[28] D. Randall, P. Goel, and R. Abujamra, "Blockchain applications and use cases in health information technology," Journal of Health & Medical Informatics, vol. 8, no. 3, 2017.

[29] M. Mettler, "Blockchain technology in healthcare: The revolution starts here," Proceedings of IEEE 18th International Conference on e-Health Networking, Applications and Services, 2016.

[30] H. Jahankhani et al. (eds.), Blockchain and Clinical Trial; Securing Patient Data. Springer, 2019.

[31] D. Metcalf et al., Blockchain in Healthcare: Innovations that Empower Patients, Connect Professionals and Improve Care. Merging Traffic Inc., 2019.

[32] P. B. Nichol, The Power of Blockchain for Healthcare: How

Blockchain Will Ignite The Future of Healthcare. Peter B. Nichol, 2017.

[33] L. Sebastian, Ultimate Blockchain Technology: Mega Edition – Six Books – Best Deal For Beginners in Blockchain, Blockchain Applications, Cryptocurrency, Bitcoin, Mining and Investing, 2019.

[34] B. K. Rai, G. Kumar, and V. Balyan(eds.), AI and Blockchain in Healthcare. Springer 2023.

[35] G. C. Deka and S. Namasudra (eds.), Applications of Blockchain in Healthcare. Springer, 2020.

[36] S. Stawicki (ed.), Blockchain in Healthcare: From Disruption to Integration. Springer, 2023.

[37] G. Thirumurugan, Blockchain Technology in Healthcare. Independently Published, 2020.

[38] M. A. Alam, P. Agarwal, and S. M. Idrees, Blockchain for Healthcare Systems: Challenges, Privacy, and Securing of Data. Boca Raton, FL: CRC Press, 2021.

[39] V. Dhillon et al., Blockchain in Healthcare. Taylor & Francis, 2021.

[40] A. Chaudhary and R. Chadha, Analyzing Blockchain in Healthcare: Applicability and Empirical Evidence of Blockchain Technology in Health Science (English Edition). BPB PUBN, 2022.

[41] G. C. Deka, M. D. Borah, and P. Zhang, Prospects of Blockchain Technology for Accelerating Scientific Advancement in Healthcare. IGI Global, 2021.

[42] B. Bhushan et al. (eds.), Blockchain Technology in Healthcare Applications: Social, Economic, and Technological Implications. Boca Raton, FL: CRC Press, 2022.

CHAPTER 5

BLOCKCHAIN IN AGRICULTURE

"Everything will be tokenized and connected by a Blockchain one day."

— Fred Ehrsam

5.1 INTRODUCTION

Agriculture is the production of food, fiber, and data. It has played a crucial role in the promotion of life and wellbeing around the world. The sector is crucial to global prosperity. The well-being of every nation is depends greatly on agriculture. Only few industries have a greater impact than agriculture. However, agriculture faces challenges of food security, food safety, sustainable development, rapidly increasing world population, consumer demand for more information and greater transparency, urbanization, and globalization. To adapt to these challenges, the agricultural sector needs to undergo technological transformation to [1]:

- Satisfy the demands of a growing population for more high-quality food
- Advance technological solutions to meet changing consumer needs
- Encourage sustainable agricultural practices and lower environmental footprints
- Decrease agricultural supply chain costs
- Establish and follow firm sanitary and phytosanitary standards
- Sustain profitable operations of farmlands and agribusinesses
- Increase the incomes of small farms, private farmers, and food producers

Today, consumers want to know exactly where their food comes from. This has resulted in agribusinesses searching for supply chain management software to improve food safety, and the traceability of the whole farming supply chain.

Finance was the first sector to which Blockchain (BC) technology was applied. In December 2016, the company AgriDigital successfully executed the world's first sale of 23.46 tons of grain on a Blockchain. It is currently used in many other areas such as healthcare, smart cities, smart contracts, energy markets, and government sector [2]. The Blockchain, when combined with big data technology, can allow multiple parties to interact cooperatively. It can help city government achieve their digital economy goals while ensuring maximum network security.

The Blockchain is a ledger of accounts and transactions that are written and stored by all participants. It refers to a digitized platform that stores and verifies transactions between users of a system. The technology offers a reliable approach of tracing transactions between anonymous participants, thereby quickly detecting fraud and malfunctions. Several companies and government agencies are experimenting with the promising technology. Blockchain has different uses in the agricultural industry such as providing solutions to food safety, food waste, food fraud, supply chain visibility, and management. Several farmers and agribusinesses have started introducing Blockchain technology in agriculture [3].

This chapter examines the applications of Blockchain technology in agriculture. It begins with providing an overview of Blockchain to make the chapter self-contained. It explains how Blockchain technology is used to improve some areas of agriculture. It covers the benefits and challenges of Blockchain in agriculture. The last section concludes with comments.

5.2 OVERVIEW OF BLOCKCHAIN

The term "Blockchain" refers to the way BC stores transaction data – in "blocks" that are linked together to form a "chain." The chain grows as the number of transactions increases. Since every entry is stored as a block on a chain, the care you receive is added to your personal ledger.

At its core, Blockchain is a distributed system recording and storing transaction records. In a Blockchain system, there is no central authority. Instead, transaction records are stored and distributed across all network participants. Rather than having a centrally located database that manages records, the database is distributed to the networks and transactions are kept secure via cryptography. BC eliminates the need for a middleman that traditionally facilitates such transactions. Unlike traditional trading systems, no intermediary is needed to track the exchange; all parties deal directly with each other [4]. Figure 5.1 illustrates how Blockchain works [5].

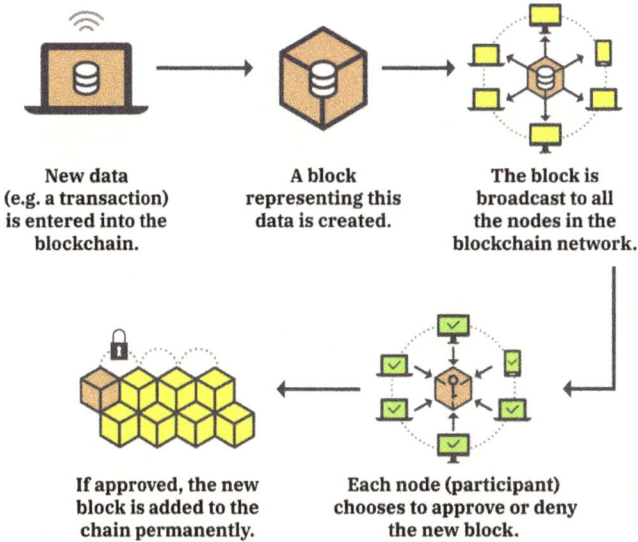

New data
(e.g. a transaction)
is entered into the
blockchain.

A block
representing this
data is created.

The block is
broadcast to all
the nodes in the
blockchain network.

If approved, the new
block is added to the
chain permanently.

Each node (participant)
chooses to approve or deny
the new block.

Figure 5.1   How Blockchain works [5].

The Blockchain was designed so transactions are immutable, i.e. they cannot be deleted. Thus, Blockchains are secure and meddle-free by design. Data can be distributed, but not copied. When it comes to digital assets and transactions, you can put almost anything on a Blockchain. Different scenarios call for different Blockchains.

The BC technology currently has the following features [6,7]:

1. *Peer-to-Peer (P2P) Network*: The first requirement of BC is a network, an infrastructure shared by multiple parties. This can be a LAN at a small scale or the Internet at a large scale. All nodes participating in a BC are connected in a decentralized P2P network. Transactions are broadcast to the P2P network. Due to some limitations of P2P networks, some vendors have provided cloud-based BCs.

2. *Cascaded Encryption*: A BC uses encryption to protect transaction data. Blocks are encrypted in a cascaded manner, i.e. the encryption result of the previous block is used in encrypting the current block. The BC is secured by public key cryptography, with each peer generating its own public-private key pairs.

3. *Distributed Database*: A BC is digitally distributed across a number of computers. Each party on a BC has access to the entire database and no single party controls the data or the information. Since BC is decentralized, there is no need for central authorizes such as banks.

4. *Transparency with Pseudonymity*: Each node or participant on a Blockchain has a unique 30-plus-character alphanumeric address that identifies it. Users can choose to remain anonymous or provide proof of their identity to others.

5. Irreversibility of Records: Once a transaction is entered in the database and the accounts are updated, the records cannot be altered. Records on the database is permanent, chronologically ordered, and available to all others on the network.

There are two types of blockhains: public and private. Public Blockchains are cryptocurrencies such as Bitcoin, enabling peer-to-peer transactions. Private Blockchains use Blockchain-based platforms such as Ethereum or Blockchain-as-a-service (BaaS) platforms running on private cloud infrastructure. A private BC is an intranet, while a public BC is the Internet. Companies will be disrupted the most by public Blockchains.

BCs may be permissioned or permissionless. In a permissioned BC, each participant has a unique identity. Permissionless BCs are appealing because they allow anyone to join, participate or leave the protocol execution without seeking permission from a centralized or distributed authority. However, permissionless BCs, such as Ethereum or Bitcoin, face transaction volume constraints [8].

There is an impressive, growing list of companies that have started to use to agriculture to safeguard food safety. For example, Walmart, Kroger, Alibaba, IBM, and other companies are implementing Blockchain food traceability projects, and using Blockchain technology to track the entire process of food production. Coca-Cola has employed it to identify cases of forced labor in the sugarcane supply chain [9]. Some of the applications of Blockchain are illustrated in Figure 5.2 [10]. Figure 5.3 shows how Blockchain works in agriculture [11].

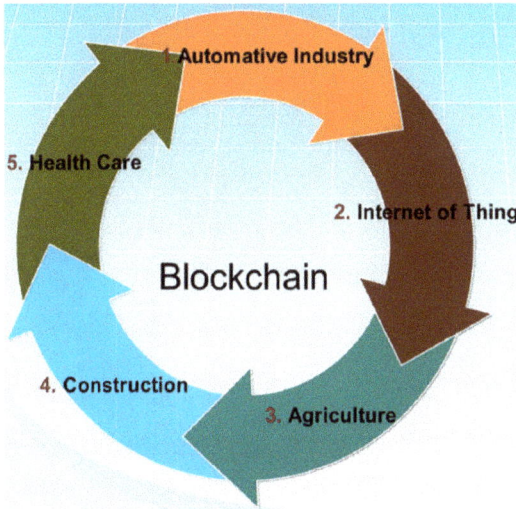

Figure 5.2 Different applications of Blockchain [10].

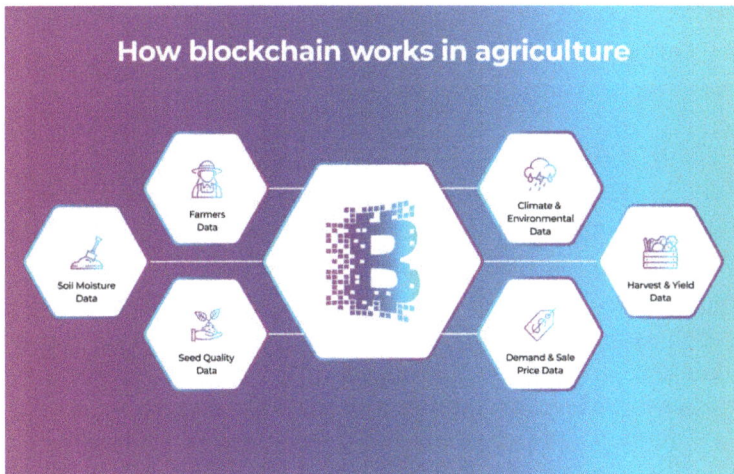

Figure 5.3   How Blockchain works in agriculture [11].

5.3 APPLICATIONS OF BLOCKCHAIN IN AGRICULTURE

Blockchain consists of an electronic system that allows electronic recordkeeping, validation, and verification without the need for an intermediary. Blockchain technology improves the following areas within the agricultural sector: crop insurance, traceability, smart farming, food supply chain, controlling weather crisis, and transactions of agricultural products [12,13]. Figure 5.4 illustrates various applications of Blockchain in agriculture [14].

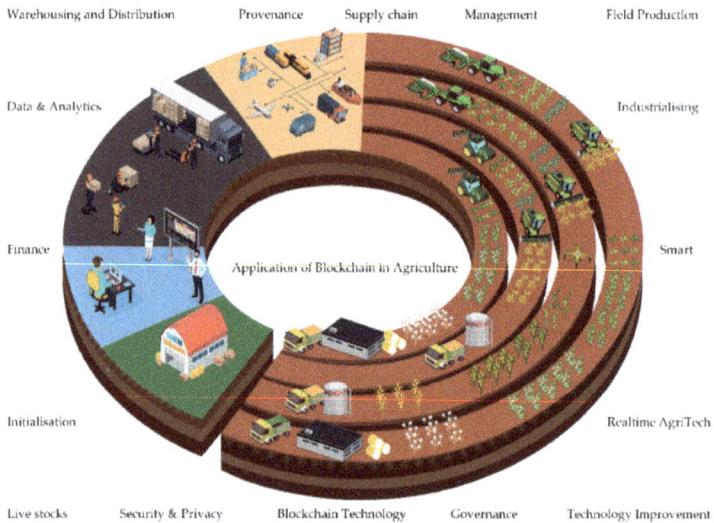

Figure 5.4 Applications of Blockchain in agriculture [14].

Some of these applications are discussed as follows.

- *Traceability*: This is the predominate area in agriculture where Blockchain is being used . The Blockchain tracking system is different from the barcode and RFID systems. Blockchain technology enables consumers to verify the journey of their product, tracing it from farm to table. It enables the traceability of information in the food supply chain and thus helps improve food safety. It also provides data on when a product was harvested in a matter of seconds. Since the information recorded on the Blockchain is unalterable, it can provide reliable information and is forgery-proof. Figure 5.5 displays traceability system product and information flow [15].

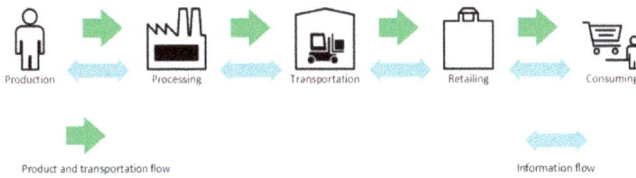

Figure 5.5   Traceability system product and information flow [15].

- *Crop Insurance*:   Weather condition can threaten agricultural production, putting food security at risk. Agricultural insurance schemes are traditionally recognized as a tool to manage weather related risks. Smart contracts help farmers insure their crops and claim damages with insurance companies. Unpredictable weather anomalies make it difficult to correctly estimate the losses they cause and this leaves room for fraud. Smart contracts insure a farmer's crops and claim damages. Using tailored smart Blockchain contracts, the damage claim can be triggered via changes to weather conditions that meet certain criteria.

- *Food Supply Chain*: Blockchain technology is a disruptive technology that changes business and supply chain models. Supply chain refers to the production and distribution processes of goods and services from suppliers to customers. The food you buy from superstores comes from different places such as Africa, Latin America, Europe, Asia, and even Alaska. It is always expedient to provide information on the origins of food products to ensure customer loyalty and confidence. With traditional supply chains, food retailers do not have an effective way of ensuring that all products were grown under specified conditions. The aim of using Blockchain technology in the food supply chain is to fulfill the desire of a traceable and transparent system. Blockchain technology allows goods and individuals to be tracked throughout the supply chain based on real time. That is why retail giants such as Walmart resort to Blockchain for tracing food products' places of origin. The time it takes to track the origin of food is drastically cut down to just 2 seconds. The Blockchain technology in the food supply chain is still in the early stages of development. Figure 5.6 shows the simplified version of the agri-food supply chain

[16].

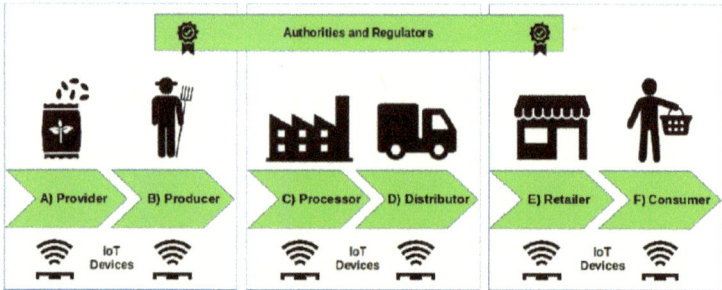

Figure 5.6   Simplified version of the agri-food supply chain [16].

• *Transactions*: Blockchain technology allows individuals and businesses to make instant network transactions without any intermediary. Application of cryptocurrency in the transaction of agricultural products will reduce transaction costs more substantially. Each user has a copy of the ledger and access to transaction information. The information of the products is recorded and controlled in these transaction records. Blockchain also allows agricultural producers to set prices more efficiently and effectively. This allows managing their output to match the demand for their products.

• *Smart Agriculture*:   Smart agriculture refers to the application of the new technologies (such as Internet of things (IoT), cloud computing, global positioning system, artificial intelligence, and big data) into traditional agriculture with the goal of reducing human effort. It implies the wise use of natural resources and the reduction of environmental impact. Someone has claimed that [17]:

Blockchain + IoT = smart agriculture

A key issue of establishing smart agriculture is developing a comprehensive security system that facilitates the use and management of data. Traditional way of managing data is a centralized fashion and is prone to inaccurate data, data distortion and misuse as well as cyber-attack. The Blockchain technology serves to store data and information that various actors and stakeholders generate throughout the entire value-added process [18].

• *Controlling Weather Crisis*: Farmers usually experience unpredictable weather conditions while growing crops. Being able to predict and monitor weather conditions is therefore essential to crop survival. Consumers do not know when the crops suffered horrible weather conditions and what led to the increased costs. Blockchain has the ability to offer traceability and transparency and provide a clear understanding of the price differences in the food distribution market. Placing agricultural weather stations within the farms can help generate crucial information such as temperature, rainfall, wind speed and direction, and atmospheric pressure. The parameters are measured, recorded, and saved in the Blockchain enabling farmers to access them transparently.

Other applications of Blockchain technology in agriculture include smart contracts, digital documents, crop and food production, management of agricultural finance, information systems, fraud prevention, automation of work, and decentralization of information.

5.4 BENEFITS
The Blockchain technology is reshaping agriculture industry's way of doing business by boosting transaction speeds, helping farmers control and analyze crops, etc. It is revolutionizing agriculture sector by enhancing the decision-making capabilities of organizations. Blockchain has the ability for agricultural producers to set prices more efficiently. Blockchain agriculture has the potential to increase efficiency, transparency, and trust throughout agricultural supply chains. The use of data distributed in the Blockchain network enables transparency of decisions along the supply chain. The stakeholders involved have access to reliable information, allowing better planning and better market control. Other benefits include [19-21]:

• *Transparency*: Blockchain can be implemented such that all transactions are stored in a distributed and transparent ledger, which makes the transaction process transparent. Consumers will be liberated from fakes and regain confidence in ecommerce. This will create trust, transparency, and confidence between producers and consumers. By utilizing Blockchain technology, producers of products can keep an accurate record of origin data, such as date and location. This method guarantees trustworthiness between involved parties. As shown in Figure 5.7, transparency refers to product attributes in the consumer value driver plate [22].

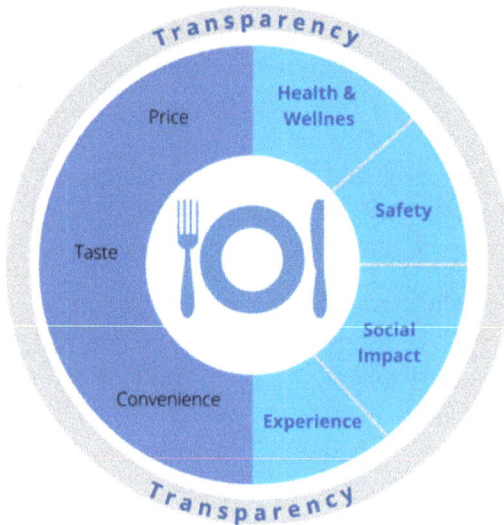

Figure 5.7   Transparency refers to product attributes [22].

- *Trust*: Blockchain technology increases the trust among consumers. This makes consumers aware of the nooks and corners of the supply chain process. They will get a transparent report about the species, origin, cultivation details, price, and so on. This builds trust between consumers and farmers and makes the relationship better and more sustainable.

- *Information Security*: Blockchain technology provides private key encryption which is a powerful tool that provides the authentication requirements. Data can only be added to a Blockchain. Once a block has been created, it cannot be altered, securing the information within it.

- *Supply Chain Management*: Blockchain technology can enable supply chain management to operate more efficiently than traditional monitoring mechanisms. Every link in the supply chain represents a "block" of information, with the advantage of visibility, aggregation, validation, automation, and resiliency.

• *Payment Methods*: The Blockchain provides a digital payment solution with zero rates. Application of cryptocurrency in the transaction of agricultural products will significantly reduce transaction costs.

• *Reduce the Cost of Farmers*: Many agricultural products are produced by households. Their low transaction volume excludes these participants from the market. Blockchain technology can greatly reduce transaction costs and incorporate them into the market again.

• *Improved Quality Control and Food Safety*: Crop failure, for example, is a prevalent challenge faced by farmers worldwide. It usually happens because of unfavorable climatic conditions, such as poorly distributed rainfall and erratic weather. Blockchain can help us to ensure optimal quality control conditions.

• *Increased Traceability*: Today, more and more consumers want to know where their food comes from. Using Blockchain technology will let consumers know exactly where their food originated, who planted it, and how fresh it is. Increasing the traceability of the supply chain will have a considerable impact on reducing food fraud and enabling consumers to know what they are paying for.

• *Increased Efficiency*: Blockchain technology allows farmers to store their data in one place and be accessed by those who need it. This simplifies the entire process, saves valuable time and energy, and increases efficiency of the farmer.

• *Fairer Payment*: Weather conditions, inelastic demand and supply, and the conditions of the global market can impact most farmers' incomes. In the current system, it often takes weeks for farmers to get paid for their goods and payment options such as wire transfers are usually costly. With a Blockchain-linked mobile store of data about transaction trends, farmers will be able to negotiate fairer prices. Blockchain-based smart contracts work by triggering payments automatically as soon as a specific, previously-specified condition has been fulfilled. This way farmers can receive payment for their goods as soon as they are delivered.

• *Decentralization*: Blockchain is a decentralized system that eliminates the intervention of mediators. In agriculture, the decentralized network plays a significant role by removing the middlemen, thus increasing security, and it helps farmers receive their sweat worth.

• *Smart Contracts*: Blockchain has a technique called smart contracts where farmers can make their transaction process more manageable and secure. Smart contracts are usually automatically generated. Due to smart contracts solutions, farmers can get their payments in time.

• *Insurance Claim*: Blockchain makes the process of insurance claims accessible. Many farmers in our country struggle for their insurance.

• *Crop Rotation*: By using the data stored in the Blockchain ledger, various data analytics techniques can be used to improve the yield. Using the data in the ledger, different crop rotation techniques can be implemented, which improves the profit for the farmers.

• *Fraud Elimination*: Blockchain does not allow any data modifications. It is accessible for everyone, providing higher protection from fraud.

5.5 CHALLENGES
In spite of the enormous benefits of Blockchain, key challenges and limitations remain for applying the Blockchain technology in agriculture. Current uncertainties are preventing individual parties from developing a convincing business case. Other challenges include [23]:

• *Complex Supply Chain*: The supply chain challenge in agriculture is getting agricultural products from farmers to consumers. There is no immediate solution for products damaged during transport by implementing Blockchain. The complexity of the modern food supply chain has created distance between the consumer and producer. Supply chain management in agriculture is more complex than other supply chains since agricultural production involves factors like weather, pests, and diseases that are hard to predict and control. Blockchain does not seamlessly integrate with existing legacy systems. The application of Blockchain technology requires wide participation and collaboration of involving parties in the agriculture sector.

- *Consumer Trust*: Trust is crucial when we desire to improve our economic interactions and impacts. Blockchain technology has changed our notion of trust. An inherent feature of the Blockchain technology is its redefining element of "trust." Fear of counterfeiting is a strong reason shoppers may bypass ecommerce platforms.

- *Legal issues*: There are some legal hurdles to clear before Blockchain can really fulfill its potential. Currently, there is no established governance system regulating Blockchain transactions.

- *Scalability*: As the number of users in a Blockchain grows, so does the number of operations. The computational power required for these operations may outpace the workload that hard disks are realistically able to handle. Current agricultural technologies cannot sustain the high transaction speeds required for the Blockchain to work, especially when it comes to large-scale projects. Issues on the number of systems required and transactions made and the volume of data to process remain.

- *Instability*: The agriculture sector is always in a financially unstable position due to several risks and factors like poor crop production due to uncertain change in climatic conditions, poor irrigation methods, etc. The biggest loss farmers face is that they get paid peanuts against the supplies they produce and sell.

- *Supply chain*: One area in agriculture that Blockchain will have difficulty improving is the supply chain. It is doubtful whether cryptocurrencies, Blockchain's payment system, can be used in food supply chains due to its volatility arising from speculation and price fluctuations.

• *Awareness*: The Blockchain itself is regarded as too complicated, making it difficult to convince farmers to integrate. There is a glaring lack of awareness, made even worse by the absence of training programs to help growers understand the technology and consequently invest. Most farmers may not be adept in using technology.

• *Regulations*: All things considered, Blockchain will be effective if it can be easily understood and used. But the government must come up with clear regulations to encourage adoption.

• *Energy Consumption*: Blockchain consumes a gargantuan amount of power when running on a grand scale. Having several network users competing to validate the same operations is a huge waste of energy that has a strikingly negative impact on the environment.

5.6 CONCLUSION

Blockchain is a distributed ledger of transactions that stores encrypted information into a chain of blocks. It has some unique features such as immutability and transparency, which disallow any fraudulent modifications to the data. The Blockchain technology is under development to support agricultural finance by many financial institutions and commercial banks. Blockchain is a disruptive technology that changes businesses and supply chains. It can provide an innovative solution for product traceability in agriculture and food supply chains. The application of Blockchain technology in agriculture is still in its infancy. Blockchain is poised to be the disruptive force that propels the agricultural industry into the 1st century.

Blockchain technology is yet to revolutionize the agriculture sector the way it has taken the financial services sector by storm. It is still new and experimental, and faces a number of significant barriers to adoption in agriculture due to some shortcomings and socio-economic challenges. The future for Blockchain in agriculture is still unclear.

More information about Blockchain in agriculture can be found in the books in [24-32] and the following related journals:

Frontiers in Blockchain and Journal of the Science of Food and Agriculture.

REFERENCES

[1] "How to apply Blockchain for supply chain in agriculture," January 2020,

https://www.intellias.com/how-to-apply-the-Blockchain-to-agricultural-supply-chains-while-avoiding-embarrassing-mistakes/

[2] M. N. O. Sadiku, Y. Wang, S. Cui, and S. M. Musa, "A primer on Blockchain," International Journal of Advances in Scientific Research and Engineering, vol. 4, no. 2, February 2018, pp. 40-44.

[3] M. N. O. Sadiku, K. G. Eze, and S.M. Musa, "Blockchain in agriculture," Information and Security, vol. 49, 2021, pp. 154-155.

[4] M. Winterson, "Opinion: Is Blockchain the breakthrough smart cities need?" September 2018,

https://www.arabianbusiness.com/technology/405261-abe-1935-is-Blockchain-the-breakthrough-smart-cities-need

[5] G. O. Rodriguez, "What is Blockchain?" June 2022,

https://money.com/what-is-Blockchain/

[6] M. Iansiti and K. R. Lakhani, "The truth about Blockchain," Harvard Business Review, Jan./Feb. 2017.

https://hbr.org/2017/01/the-truth-about-Blockchain

[7] W. T. Tsai et al., "A system view of financial Blockchains," Proceedings of IEEE Symposium on Service-Oriented System Engineering, 2016, pp. 450-457.

[8] Z. Alhadhrami et al., "Introducing Blockchains for healthcare," Proceedings of

International Conference on Electrical and Computing Technologies and Applications, 2017.

[9] A. Kamilaris , F. X. Prenafeta-Boldú, and A. Fonts, "The rise of Blockchain technology in agriculture," September 2018,

https://ictupdate.cta.int/en/article/the-rise-of-Blockchain-technology-in-agriculture-sid054973f61-0cc3-4a9c-914a-f3a27ecd0f52

[10] S. V. Akram et al., "Adoption of Blockchain technology in various realms: Opportunities and challenges," Security Privacy, vol. 3, 2020.

[11] "Blockchain in agriculture: The four progressive application cases,"

https://www.towardsanalytic.com/Blockchain-in-agriculture-the-four-progressive-application-cases/

[12] "8 Blockchain startups disrupting the agricultural industry,"

https://www.startus-insights.com/innovators-guide/8-Blockchain-startups-disrupting-the-agricultural-industry/

[13] M. H. Ronaghi, "A Blockchain maturity model in agricultural supply chain,"

Information Processing in Agriculture, November 2020.

[14] L. B. Krithika, "Survey on the applications of Blockchain in agriculture," Agriculture, vol. 12, 2022.

[15] K. Demestichas et al., "Blockchain in agriculture traceability systems: A review,"

Applied Sciences, vol. 10, 2020.

[16] M. P. Caro et al., "Blockchain-based traceability in agri-food supply chain management: A practical implementation," Proceedings of IoT Vertical and Topical Summit on Agriculture, Tuscany, May 2018.

[17] "Blockchain in agriculture: Benefits, use cases and platforms that are disrupting the industry," January 2021,

https://pixelplex.io/blog/Blockchain-in-agriculture/

[18] M. Torky and A. EllaHassanein, "Integrating Blockchain and the internet of things in precision agriculture: Analysis, opportunities, and challenges," Computers and Electronics in Agriculture, vol. 178, November 2020.

[19] H. Xiong et al., "Blockchain technology for agriculture: Applications and rationale,"

Frontiers in Blockchain, February 2020.

[20] A. Takyar, "Blockchain in agriculture – Improving agricultural techniques"

https://www.leewayhertz.com/Blockchain-in-agriculture/#:~:text=Blockchain%20brings%20fairness%20in%20the,transparency%20and%20shared%20control%20 accessibility.&text=Every%20time%20a%20transaction%20 will,to%20access%20every%20transaction%20transparently.

[21] "Top 10 benefits of Blockchain in agriculture,"

https://www.techyv.com/article/top-10-benefits-of-Blockchain-in-agriculture/

[22] "Blockchain in agriculture empowering brands & consumers," May 2021,

https://tracextech.com/Blockchain-in-agriculture-making-impact/

[23] "Blockchain in agriculture: Getting it off the ground," November 2020,

https://www.techslang.com/Blockchain-agriculture-getting-it-off-the-ground/

[24] International Telecommunication Union, Food and Agriculture Organization of the United Nations, E-agriculture in Action Blockchain for Agriculture : Opportunities and Challenges. International Telecommunication Union, 2015.

[25] R. A. Hadiguna et al., Agriculture Blockchain. Vital Wellspring Education Pte. Ltd., 2022.

[26] X. Wang, Blockchain Chicken Farm: And Other Stories of Tech in China's Countryside. Farrar, Straus and Giroux, 2020.

[27] C. Fennell, Visioning the Agriculture Blockchain: The Role and Rise of Blockchain in the Commercial Poultry Industry. Michigan State University. Information and Media, 2022.

[28] T. Tomu, The Rise of Blockchain for Agriculture: Exploring the Opportunities, Benefits, Limitations and Risks Associated with Applying Distributed Ledger Technology to Agriculture. Kindle Edition, 2020.

[29] S. Kim and G. Deka G. (eds), Advanced Applications of Blockchain Technology. Singapore: Springer, 2020.

[30] K. Cottrill, S. Decovny, and P. Harris, Blockchain and the Future of Food: Driving Efficiency, Transparency and Trust in Food Supply Chains. Unknown Publisher, 2018.

[31] L. Ge et al., Blockchain for Agriculture and Food: Findings from the Pilot Study.

Wageningen Economic Research, 2017.

[32] V. Wassenaer et al., Applying Blockchain for Climate Action in Agriculture: State of Play And Outlook. Food and Agriculture Organization of the United Nations, 2021.

CHAPTER 6

BLOCKCHAIN IN SUPPLY CHAIN

"Blockchain delivers trust between different parties exchanging data and this is why we use Blockchain for."

— Anonymous

6.1 INTRODUCTION

Over the centuries, the preservation of valuable assets has made it necessary to store them in a physical space. A secure storage space has been generated with the Blockchain technology. Blockchain is a distributed ledger that does not have a single record database in which all Bitcoin operations are performed. Blockchain's database is the most important feature that makes it safe to be distributed, as the community agrees in all actions through this. The application of Blockchain to the supply chain has attracted the attention of many business owners as it can be quickly adapted to dynamic market conditions and in the business environment.

Processes associated with the transformation of inputs (raw materials, components) into outputs (goods, finished products) and their transport to the place of consumption are an essential part of the modern society. Supply chain plays a critical role in the global economy. As illustrated in Figure 6.1, supply chain is a set of sequential stages in the manufacturing, transportation, storing, or distribution of a product [1].

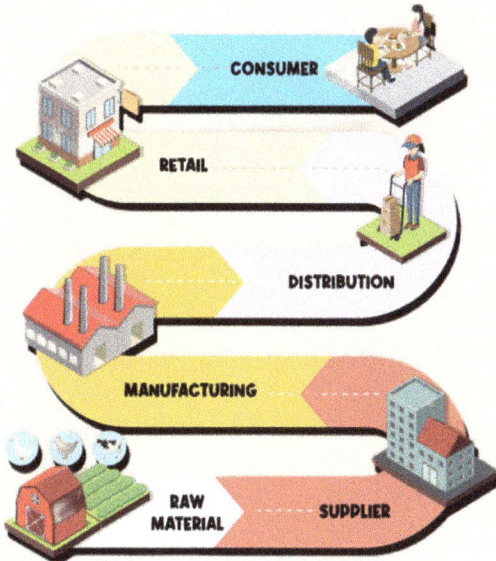

Figure 6.1   A supply chain [1].

It involves various participants and stakeholders and numerous processes in multiple stages. It serves as a vital driver for businesses aiming to achieve success by optimizing their operations, curtailing expenses, and delivering an unparalleled level of customer satisfaction. Since supply chains are, at their core, a network of interlinked companies, each business adds value to a product or service before it reaches the end user. Modern supply chain systems are managed using software from the creation of goods and services, warehousing, inventory management, order fulfillment, information tracking and product/service delivery to after-sales services. Today's supply chains are global networks that generally include manufacturers, suppliers, logistics companies, and retailers that work together to deliver products to consumers. They need to be more reliable than ever because disruptions in the supply chain can create significant losses for companies and increase costs for end customers [2,3]. The information and material flow in supply chain management is shown in Figure 6.2 [4].

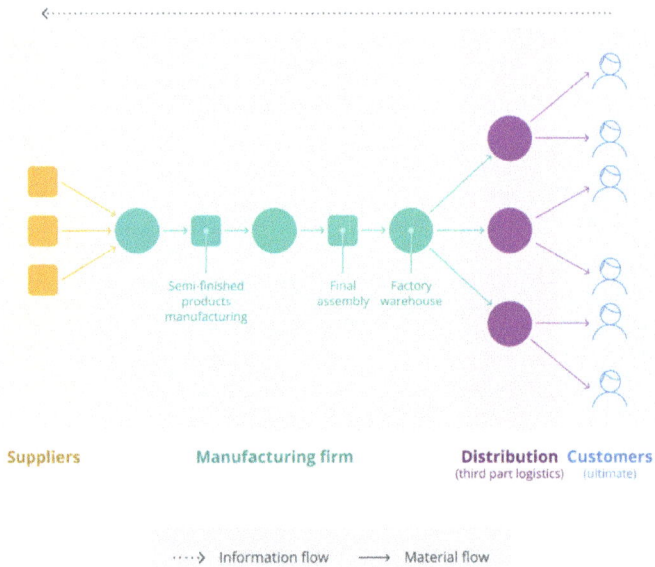

Figure 6.2 *The information and material flow in supply chain management [4].*

Blockchain technology is an innovation which is regarded as the center of Industry 4.0 revolution and it has become part of our lives. It is a system that stores data in a special way. In most cases, today's supply chains operate at-scale without Blockchain technology. Although this technology finds its first application in the financial sector, it has become possible to use it in all sectors which can be integrated with technology today [5]. The Blockchain technology has excited the IT and supply-chain worlds.

In this chapter, we introduce the integration of Blockchain in supply chain. It begins with presenting an overview on Blockchain to make the chapter self-contained. It covers some applications of BC in supply chain. It highlights the benefits and challenges of BC in supply. The last section concludes with comments.

6.2 OVERVIEW OF BLOCKCHAIN

Blockchain (BC) technology is a permanent record of online transactions. It is a distributed tamper-proof database, shared, and maintained by multiple parties. It is a new enabling technology that is expected to revolutionize many industries, including business. It has the potential for addressing significant business issues. The BC technology allows participants to move data in real-time, without exposing the channels to theft, forgery, and malice.

The term "Blockchain" refers to the way BC stores transaction data – in "blocks" that are linked together to form a "chain." The chain grows as the number of transactions increases. Since every entry is stored as a block on a chain, the care you receive is added to your personal ledger. The first Blockchain was conceived in 2008 by an anonymous person or group known as Satoshi Nakamoto, who published a white paper introducing the concept of a peer-to-peer electronic cash system he called Bitcoin [6]. Blockchain is a distributed ledger database that consists of records or transactions or various digital incidents that are executed by the participants. Blockchain technology can assist in achieving the seven objectives of SCM: their cost, quality, speed, dependency, risk reduction, sustainability as well as flexibility. The concept of Blockchain is shown in Figure 6.3 [7].

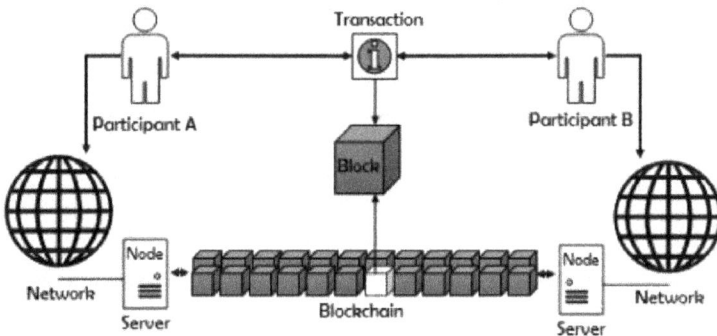

Figure 6.3   The concept of Blockchain [7].

At its core, Blockchain is a distributed system recording and storing transaction records. In a Blockchain system, there is no central authority. Instead, transaction records are stored and distributed across all network participants. Rather than having a centrally located database that manages records, the database is distributed to the networks and transactions are kept secure via cryptography. BC eliminates the need for a middleman that traditionally may facilitate such transactions.

Fundamentally, Blockchains are distributed digital database that record and maintain a list of transactions taking place in real time. They may also be regarded as decentralized ledgers that sequentially record transactions or interactions among users within a distributed network. They have the following properties [8]:

- Firstly, they are autonomous. They run on their own, without any person or company in charge.

- Secondly, they are permanent. They are like global computers with 100 percent uptime. Because the contents of the database are copied across thousands of computers, if 99 per cent of the computers running it were taken offline, the records would remain accessible and the network could rebuild itself.

- Thirdly, they are secure and tamper-proof. Each record in Blockchain is time stamped and stored cryptographically. The encryption used on Blockchains like Bitcoin and Ethereum is industry standard, open source, and has never been broken.

- Fourthly, they are open, allowing anyone to develop products and services on them.

- Fifthly, as Blockchain is a shared system, costs are also shared between all of its users.

The Blockchain was designed so transactions are immutable, i.e. they cannot be deleted. Thus, Blockchains are secure and meddle-free by design. Data can be distributed, but not copied. When it comes to digital assets and transactions, you can put almost anything on a Blockchain. Different scenarios call for different Blockchains. Blockchain is used for different purposes as depicted in Figure 6.4 [9].

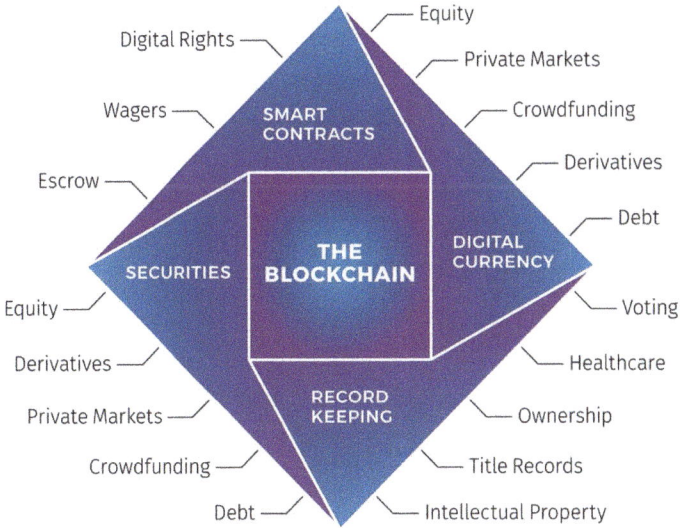

Figure 6.4   Different purposes of Blockchain [9].

The BC technology currently has the following features [10,11]:

1. *Peer-to-Peer (P2P) Network*: The first requirement of BC is a network, an infrastructure shared by multiple parties. This can be a LAN at a small scale or the Internet at a large scale. All nodes participating in a BC are connected in a decentralized P2P network. Transactions are broadcast to the P2P network. Due to some limitations of P2P networks, some vendors have provided cloud-based BCs.

2. *Cascaded Encryption*: A BC uses encryption to protect transaction data. Blocks are encrypted in a cascaded manner, i.e. the encryption result of the previous block is used in encrypting the current block. The BC is secured by public key cryptography, with each peer generating its own public-private key pairs.

3. *Distributed Database*: A BC is digitally distributed across a number of computers. Each party on a BC has access to the entire database and no single party controls the data or the information. Since BC is decentralized, there is no need for central authorizes such as banks.

4. *Transparency with Pseudonymity*: Each node or participant on a Blockchain has a unique 30-plus-character alphanumeric address that identifies it. Users can choose to remain anonymous or provide proof of their identity to others.

5. *Irreversibility of Records*: Once a transaction is entered in the database and the accounts are updated, the records cannot be altered. Records on the database is permanent, chronologically ordered, and available to all others on the network.

As illustrated in Figure 6.5, a Blockchain comprises a peer-to-peer network of participant nodes, a distributed ledger consisting of immutable blocks of data, transactions recorded in the blocks, smart contracts to execute the transactions, and a consensus algorithm that decides the proposer of the next block [1].

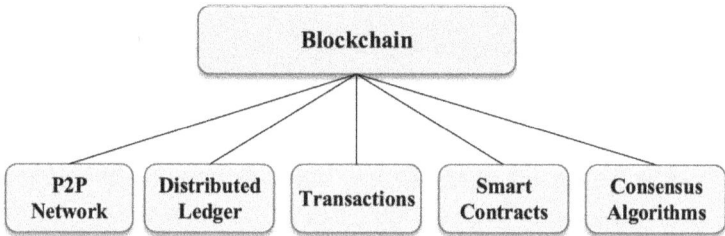

Figure 6.5 Components of a Blockchain [1].

There are two types of Blockchains: public and private. Public Blockchains are cryptocurrencies such as Bitcoin, enabling peer-to-peer transactions. Private Blockchains use Blockchain-based platforms such as Ethereum or Blockchain-as-a-service (BaaS) platforms running on private cloud infrastructure. A private BC is an intranet, while a public BC is the Internet. Companies will be disrupted the most by public Blockchains.

6.3 APPLICATIONS

Blockchain technology (BCT) has also received considerable attention outside the financial sector. Over the past few years, the Blockchain concept has attracted many industries. Several companies have identified possible use cases that could benefit from Blockchain. Figure 6.6 shows participants and their roles in a typical Blockchain integrated supply chain flow [12].

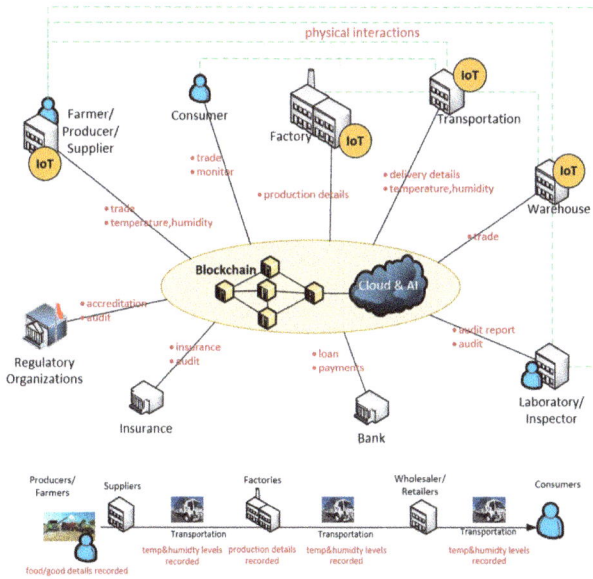

Figure 6.6 Participants and their roles in a typical Blockchain integrated supply chain flow [12].

Several prominent, forward-thinking companies are testing Blockchain solutions and investigating Blockchain uses for their supply chains. Several applications of Blockchain technology in supply chain are discussed in business practice. Some of these applications include the following [1,13-16]:

- *Supply Chain Management*: This refers to the coordination and management of activities involved in the production and delivery of products and services. Figure 6.7 shows supply chain management (SCM) flow [17].

Figure 6.7 Supply chain management (SCM) flow [17].

SCM generally involves planning, information, source, manufacturing, inventory, production, location, transportation, and return of goods. Blockchain technology can be integrated into supply chain management (SCM) to enhance transparency, traceability, and security. In the supply chain, Blockchain is best for tracking the movement of goods as they change from supplier to manufacturer to retailer and finally to consumer. SCM refers to controlling the entire production flow, from acquiring raw materials to delivering the final product at the destination. In addition, it handles the movement of materials, information and finances associated with a good or service. Traditional SCMs involve steps like planning, sourcing, manufacturing, delivering and after-sales service to control the supply chain centrally. Blockchain harmonizes communication systems across the supply chain onto a unified platform. The utilization of Blockchain in supply chain management necessitates adherence to regulatory requirements.

• *Logistics*: Supply chain refers to the system of organizations, people, activities, information, and resources involved in moving a product or service from supplier to customer. Logistics is only one part of the supply chain. Logistics and supply chain management are regarded as domains where Blockchains are good fits for many reasons. The integration of smart contracts within Blockchain technology offers the ability to autonomously verify, record, and coordinate transactions without the need for third-party intermediaries. At the moment, BC technology is still far from maturity with many challenges to overcome before it can be successfully deployed at scale in the logistics industry.

• *Food Supply Chains*: Because food is essential, food supply chains have attracted more interest in applying Blockchain technology than any other industry. The food industry is actively exploring the potential of Blockchain technology to enhance the safety and integrity of the food supply chain. The idea of being able to trace ingredients of any food or product back to its origin is very compelling. Food supply chains comprise many stages and they may not be finely monitored and tracked. As a result, end consumers are usually unable to trace their food products' origins. For the most part, the food we eat is the result of a complex global supply chain, comprising a complex web of production, processing, packaging, storage and distribution. In 2017, several leaders in the food industry came together to figure out ways Blockchain can be used to improve global food safety. And today, companies like Unilever, Nestlé, and Walmart are using Blockchain technology. Walmart has long been known as a leader in supply chain management, and has taken a serious interest in Blockchain technology.

• *Counterfeit Drugs*: These are a common challenge for the pharmaceutical industry. Recent studies in this industry show the effectiveness of Blockchain in tracking and authenticating drugs. The pharmaceutical industry is using Blockchain to track and verify the authenticity of drugs throughout the supply chain, from production to distribution to end-users. BC reduces counterfeit medicines. It minimizes patient risk by reacting quickly to medication recalls, and reduce overall pharmaceutical costs.

• *Entertainment and Media*: This industry faces transparency challenges in its supply chain because stakeholders need to be assured of the quality of service and compliance of regulations. The complexity of this industry is due to not only the size of operations but also to the vast number of regulations.

• *Postage*: This is an industry that considers Blockchain for its supply chain to detect counterfeit stamps as the number of fraud cases continues to increase. Adversaries may take advantage of the variation of currency and counterfeit old stamps. The lack of expiration date of stamps increases the complexity of the challenge.

6.4 BENEFITS

Blockchain in the supply chain is becoming more prevalent as the benefits of the technology are realized by more and more businesses. The Blockchain technology has the potential to unlock significant value for organizations by reducing supply chain risk, increasing visibility, and enhancing trust across a complex ecosystem. The supply chain can significantly reduce the risk of unethical sourcing, shipping delays, inadequate storage, or ineffective distribution of your goods, The distributed ledger technology eliminates the need for intermediaries disrupting the ownership model. As shown in Figure 6.8, Blockchain plays a crucial role in addressing supply chain issues transparency, traceability, and trust [7].

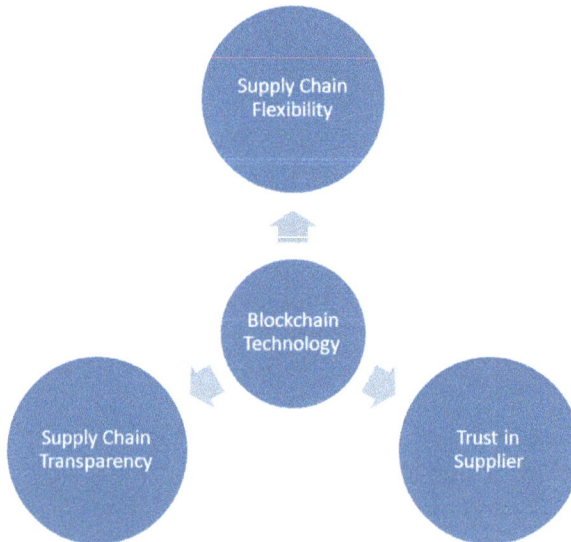

Figure 6.8 Blockchain utilization in supply chain [7].

Another benefit that Blockchain brings to the supply chain is increased data veracity and security. Other benefits of Blockchain in supply chain include the following [18,19]:

- *Enhance*: Transparency: Blockchain technology can provide real-time visibility and tracking of goods and products throughout the entire supply chain, from production to distribution to end consumers. Blockchain enhances supply chain management through process tracking, regulatory compliance, and reporting. This helps to increase transparency and trust between different parties in the supply chain. The implementation of Blockchains will bring traceability, transparency, and accountability to the movement of goods and commodities.

- *Improve Efficiency*: Blockchain can be used to track and manage the flow of goods and information throughout the supply chain. By creating a shared, immutable record of all transactions, Blockchain can help streamline supply chain management and improve efficiencies. Having an immutable, reliable record accessible to all supply chain actors can reduce the need for paperwork and significantly improve the efficiency of the entire operation.

- *Reduce Errors*: The Blockchain is a distributed database allowing secure, transparent, and tamper-proof record-keeping. This makes it an ideal solution for supply chain management as it can help reduce errors and improve transparency.

- *Improve Finance*: The cost of goods and services is always a top concern for businesses. The distributed ledger technology of Blockchain has the potential to streamline supply chains and improve finance.

- *Better Collaboration*: Blockchain enables all participants in the supply chain to have access to the same information, which can lead to better collaboration and coordination.

- *Traceability*: Tracing the activities along the supply chain allows concerned parties to access price, date, origin, quality, certification, destination, and other pertinent information using Blockchain. This improves operational efficiency by mapping and visualizing enterprise supply chains. Blockchain helps organizations understand their supply chain and engage consumers with real, verifiable, and immutable data.

- *Tradeability*: This is a unique Blockchain offering that redefines the conventional marketplace concept. Similar to how a stock exchange allows trading of a company's shares, this fractional ownership allows tokens to represent the value of a shareholder's stake of a given object. These tokens are tradeable, and users can transfer ownership without the physical asset changing hands. Recalls become less expensive and more efficient when manufacturers can locate affected products quickly and easily. Blockchain has the unique potential to track ownership records for real estate, automobiles, and digital assets.

• *Trust*: Because data on the Blockchain is decentralized and immutable, members of the supply chain can trust the data they see on the Blockchain.

• *Sustainability*: Blockchain is useful for establishing sustainable operations in supply chains. It can potentially improve the environment and enhance sustainability substantially.

• *Automation*: Automated compliance and reporting will reduce friction, reporting costs, and eliminate errors associated with manual activities.

• *Quality Control*: Blockchain can be used to track the quality of products as they move through the supply chain, reducing waste and improving customer satisfaction.

• *Finance*: Blockchain can be used to facilitate supply chain finance, providing secure and transparent records of transactions between suppliers, manufacturers, and distributors.

• *Smart Contracts*: These are self-executing contracts with the terms of the agreement between buyer and seller being directly written into lines of code. They can be implemented using Blockchain technology to automate and streamline supply chain processes, reducing costs and improving efficiency. Figure 6.9 shows a smart contract [4].

Figure 6.9 A smart contract [4].

- *Reducing Paperwork*: By digitizing and automating supply chain processes using Blockchain technology, administrative costs can be reduced, and paperwork can be eliminated, saving time and resources.

- *Food Safety*: Blockchain can be used to track and monitor food safety, ensuring that food products are safe for consumption. For example, IBM Food Trust is for food supply chain management that enables food producers, distributors, and retailers to track the movement of food products across the supply chain.

- *Compliance*: Blockchain can be used to ensure compliance with regulatory requirements by providing a transparent and auditable record of supply chain processes.

Some of these benefits are displayed in Figure 6.10 [20].

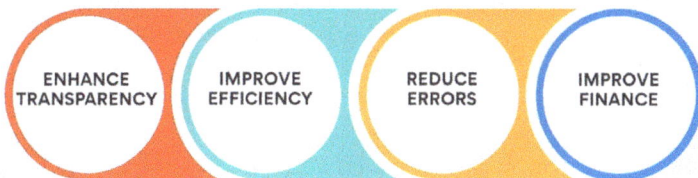

Figure 6.10 Some of these benefits of BC in supply chain [20].

6.5 CHALLENGES

Although Blockchain supports supply chain management by resolving some of the existing concerns, like provenance and quality assurance, it also introduces additional costs for implementation and maintenance and also some security concern. Some challenges remain for making the supply chain more efficient, reliable, and secure. Some of the addressed challenges are transparency, provenance, performance improvement, quality assurance and control, and achieving sustainability. Other challenges include the following [21]:

- *Complexity*: Supply chains contain complex networks of suppliers, manufacturers, distributors, retailers, auditors, and consumers. They are becoming more complex with the gradual globalization. As the complexity of the supply chain increases, so does the risk of disruption. It can be highly complex because it may comprise a large number of stages and all the involved parties need to keep track of the development of the product at each stage. The complexity of supply chain management has increased by not only directing the flow of goods but also the flow of information.

- *Loss of Transparency*: With a large number of documents, the risk of fraud and forgery has increased and the level of transparency is being lost. The multi-party nature of supply chain tends to result in quite a few inefficiencies and opportunities for fraud.

- *Privacy and Security*: These issues on Blockchains are important features for the logistics industry as they concern the information of the products, the consumers, and the interactions between them.

- *Permissioned Blockchains*: Because supply chain information can be sensitive, a permissioned Blockchain (that is, a Blockchain that is not open to the public) is usually preferred. However, a permissioned system is less secure, because there are fewer nodes to make up the Blockchain and those nodes are typically known to each other, resulting in an easier ability to change a block.

- *Human Element*: While there is great value in all members of a supply chain knowing that the data on the Blockchain cannot be changed once it is established, there can still be human error or intentional misconduct in inputting the initial data onto the Blockchain.

• *Upfront Costs*: The upfront costs of implementing a Blockchain solution in supply chain have the potential to be steep. There are costs associated with hiring Blockchain developers, which tend to cost more than traditional developers due to their specialized area of expertise. Planning costs, licensing costs, and maintenance costs can also contribute to a hefty price tag.

Despite these challenges, Blockchain technology is useful as the driver of digitization in the supply chain.

6.6 CONCLUSION

Blockchain is an Internet-based technology that is prized for its ability to publicly validate, record, and distribute transactions in immutable, encrypted ledgers. It has emerged as a promising approach to ensuring the traceability and integrity of data. By verifying and adding data in real time, Blockchain can increase transparency across a supply chain. The ability to deploy Blockchain technologies to create the next generation of digital supply chain networks will be a key element in business success.

Since Blockchain technology is still in its infancy, it is governed by various laws in many nations. It is integrated in all sectors and is applied in the supply chain. There will be rapid improvements to the technology in the near future. Forward-thinking companies are now investigating Blockchain as a technology that could potentially revolutionize supply chain as we know it. More information about the utilization of Blockchain in supply chain can be found in the books in [22-31].

REFERENCES

[1] "Blockchain in supply chain- challenges, benefits, use-cases & considerations," June 2023,

https://www.debutinfotech.com/blog/Blockchain-in-supply-chain-challenges-benefits-use-cases-considerations

[2] M. N. O. Sadiku, C. M. M. Kotteti, and J. O. Sadiku, "Blockchain in supply chain," International Journal of Advances in Scientific Research and Engineering, vol. 9, no. 9, September 2023, pp. 10-17.

[3] "How Blockchain enhances supply chain management: A survey," IEEE Open Journal of the Computer Society, vol. 1, 2020, pp. 230-249.

[4] "How Blockchain technology is used in supply chain management?"

https://cointelegraph.com/explained/how-Blockchain-technology-is-used-in-supply-chain-management

[5] S. Kaya et al., "Blockchain technology in supply chain," The Journal of International Scientific Researches, vol. 4, no. 2, 2019, pp. 121-135.

[6] M. N. O. Sadiku, Y. Wang, S. Cui, and S. M. Musa, "A primer on Blockchain," International Journal of Advances in Scientific Research and Engineering, vol. 4, no. 2, February 2018, pp. 40-44.

[7] C. Singh, R. Thakkar, and J. Warraich, "Blockchain in supply chain management," European Journal of Engineering and Technology Research, vol. 7,no. 5, October 2022, pp. 60-70.

[8] S. Depolo, "Why you should care about Blockchains: the non-financial uses of Blockchain technology," March 2016,

https://www.nesta.org.uk/blog/why-you-should-care-about-Blockchains-non-financial-uses-Blockchain-technology

[9] "Benefits of Blockchain: A business sector perspective,"

https://www.e-zigurat.com/innovation-school/blog/benefits-of-Blockchain/

[10] M. Iansiti and K. R. Lakhani, "The truth about Blockchain," Harvard Business Review, Jan./Feb. 2017.

https://hbr.org/2017/01/the-truth-about-Blockchain

[11] W. T. Tsai et al., "A system view of financial Blockchains," Proceedings of IEEE Symposium on Service-Oriented System Engineering, 2016, pp. 450-457.

[12] T. Dursun et al., "Blockchain technology for supply chain management," January 2022,

https://www.researchgate.net/publication/353764416_Blockchain_ Technology_for_Supply_Chain_Management

[13] E. Glover, " 5 Uses of Blockchain in the supply chain," September 2022,

https://builtin.com/Blockchain/Blockchain-in-supply-chain

[14] G. Blossey, J. Eisenhardt, and G.J. Hahn, "Blockchain technology in supply chain management: An application perspective," Proceedings of the 52nd Hawaii International Conference on System Sciences, 2019.

[15] M. N. O. Sadiku, K. G. Eze, and S. M. Musa, "Supply chain management," International Journal of Engineering Research, vol. 7, no. 8, August 2018, pp. 137-139.

[16] M. N. O. Sadiku, O. D. Olaleye, and S. M. Musa, "Global supply chain management," International Journal of Trend in Research and Development, vol. 7, no. 5, October 2020, pp. 11-14.

[17] M. D. Watson, " How blockchain development is transforming supply chain management,"

https://mobileappcircular.com/how-blockchain-development-is-transforming-supply-chain-management-ab7aabc2538

[18] S. Ashcroft, "Top 10 uses of Blockchain in supply chain," March 2023,

https://supplychaindigital.com/top10/top-10-uses-of-Blockchain-in-supply-chain

[19] "Blockchain in supply chain management,"

https://consensys.net/Blockchain-use-cases/supply-chain-management/

[20] "What is Blockchain in supply chain? An overview," November 2022,

https://www.edureka.co/blog/Blockchain-in-supply-chain/

[21] E. Wang and K. E. Wegrzyn, "Blockchain in supply chain series,"

https://www.foley.com/-/media/files/insights/publications/2022/06/foley-Blockchain-in-supply-chain-ebook.pdf?la=en

[22] N. Kshetri, Blockchain and Supply Chain Management. Elsevier, 2021.

[23] M. Shahbaz and M .S. Mubarik, Blockchain Driven Supply Chain Management: A Multi-dimensional Perspective. Springer Nature Singapore, 2023.

[24] N. Subramanian, A. Chaudhuri, and Y. Kayıkcı, Blockchain and Supply Chain Logistics: Evolutionary Case Studies. Springer, 2020.

[25] D. Tapscott (ed.), Supply Chain Revolution: How Blockchain Technology Is Transforming the Global Flow of Assets. Barlow Book Publishing, 2020.

[26] N. Vyas, A. Beije, and B. Krishnamachari, Blockchain and the Supply Chain: Concepts, Strategies and Practical Applications. Kogan Page, Limited, 2022.

[27] E. Hofmann, U. M. Strewe, and N. Bosia, Supply Chain Finance and Blockchain Technology: The Case of Reverse Securitisation. Springer, 2017.

[28] T. Clohessy (ed.), Blockchain in Supply Chain Digital Transformation. Boca Raton, FL: CRC Press, 2023.

[29] R. I. van Hoek et al., Integrating Blockchain Into Supply Chain Management:A Toolkit for Practical Implementation. Kogan Page,2019.

[30] N. Hackius, Blockchain Adoption in Supply Chain Management and Logistics. Tredition, 2022.

[31] A. Bouras, B. Aouni, and I. Khalil (eds.), Blockchain Driven Supply Chains and Enterprise Information Systems. Springer, 2022.

CHAPTER 7

BLOCKCHAIN IN GOVERNMENT

"Government leaders should get up to speed on the blockchain by understanding it first and committing to exploring its potential. The revolution of blockchain is not going to happen from outside the system; it's going to happen from within the system."

— William Mougayar

7.1 INTRODUCTION

Blockchain refers to a highly secure and decentralized ledger system on which information can be stored but cannot be altered. It is a technology that builds a trustworthy service in an untrustworthy environment. It has evolved beyond cryptocurrencies to general purpose and can be used across an array of applications. It is now being implemented in various industries with government being one of them. Governments around the world are taking advantage of blockchain technology to improve the efficiency and delivery of public services [1].

A blockchain is a decentralized network consisting of data records or "blocks" that cannot be modified by a single actor. It refers to a range of general purpose technologies to exchange information and transact digital assets in distributed networks [2]. It has recently emerged as a disruptive innovation with a wide range of applications, potentially able to redesign our interactions in business, politics, and society at large. Blockchain technology has been taunted as one of the most transformative technologies of our time with the most obvious use cases being the application in the cryptocurrency space. Blockchain is the technology used by developers of cryptocurrencies, like Bitcoin, to enable exchange of financial "coins" between participants without a trusted third party to ensure the transaction. One important feature in blockchain technology is decentralization. No one computer or organization can own the chain. Instead, it is a distributed ledger via the nodes connected to the chain. Blockchain is used in many areas as illustrated in Figure 7.1 [3].

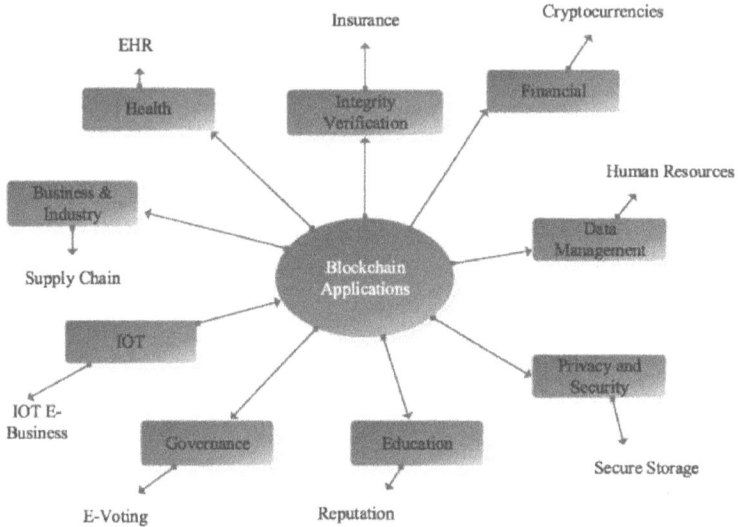

Figure 7.1 Blockchain is used in many areas [3].

Most government services around the world run on inefficient systems loaded with heavy bureaucracy. They lead to non-transparent systems and a loss of public confidence in government services. Public sectors have been taking advantage of the Internet's evolution to enhance their working processes in many ways. For example, the government has transformed itself into an electronic government to catch up with the tendency and provide efficient public services. Blockchain provides governments with a fast, secure, efficient, speedy, trustworthy, and transparent way to deliver government services and communicate with their citizens. It can help build citizens' trust, prevent data breaches, reduce corruption, and cut government spending. As a result, the governments of many countries have expressed interest in the technology. By using blockchain, governments can reduce administrative costs, increase transparency and improve service delivery. A blockchain-based government has the potential to solve legacy pain points and enable the following advantages [4]:

- Secure storage of government, citizen, and business data
- Reduction of labor-intensive processes
- Reduction of excessive costs associated with managing accountability

- Reduced potential for corruption and abuse
- Increased trust in government and online civil systems

This chapter provides an overview on blockchain in government. The chapter begins with presenting an overview on Blockchain to make the chapter self-contained. It discusses several applications of BC in government that are both representative and meaningful. It highlights the benefits and challenges of BC in government. The last section concludes with comments.

7.2 OVERVIEW OF BLOCKCHAIN

Blockchain (BC) technology is a permanent record of online transactions. It is a distributed tamper-proof database, shared, and maintained by multiple parties. It is a new enabling technology that is expected to revolutionize many industries, including business. It has the potential for addressing significant business issues. The BC technology allows participants to move data in real-time, without exposing the channels to theft, forgery, and malice.

The term "blockchain" refers to the way BC stores transaction data – in "blocks" that are linked together to form a "chain." The chain grows as the number of transactions increases. Since every entry is stored as a block on a chain, the care you receive is added to your personal ledger. The first Blockchain was conceived in 2008 by an anonymous person or group known as Satoshi Nakamoto, who published a white paper introducing the concept of a peer-to-peer electronic cash system he called Bitcoin [5]. Blockchain is a distributed ledger database that consists of records or transactions or various digital incidents that are executed by the participants. Blockchain technology can assist in achieving the seven objectives of SCM: their cost, quality, speed, dependency, risk reduction, sustainability as well as flexibility. The concept of blockchain is shown in Figure 7.2 [6].

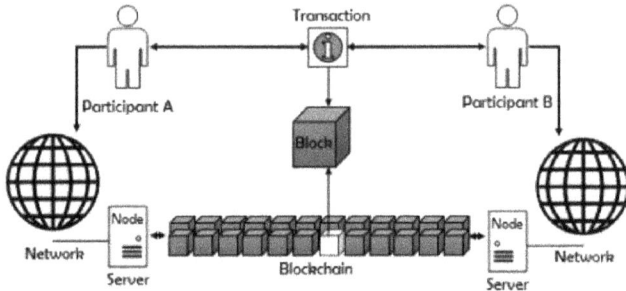

Figure 7.2 The concept of blockchain [6].

At its core, blockchain is a distributed system recording and storing transaction records. In a blockchain system, there is no central authority. Instead, transaction records are stored and distributed across all network participants. Rather than having a centrally located database that manages records, the database is distributed to the networks and transactions are kept secure via cryptography. BC eliminates the need for a middleman that traditionally may facilitate such transactions.

Fundamentally, blockchains are distributed digital database that record and maintain a list of transactions taking place in real time. They may also be regarded as decentralized ledgers that sequentially record transactions or interactions among users within a distributed network. They have the following properties [7]:

- Firstly, they are autonomous. They run on their own, without any person or company in charge.

- Secondly, they are permanent. They are like global computers with 100 percent uptime. Because the contents of the database are copied across thousands of computers, if 99 per cent of the computers running it were taken offline, the records would remain accessible and the network could rebuild itself.

- Thirdly, they are secure and tamper-proof. Each record in blockchain is time stamped and stored cryptographically. The encryption used on blockchains like Bitcoin and Ethereum is industry standard, open source, and has never been broken.

- Fourthly, they are open, allowing anyone to develop

products and services on them.

- Fifthly, as blockchain is a shared system, costs are also shared between all of its users.

The Blockchain was designed so transactions are immutable, i.e. they cannot be deleted. Thus, blockchains are secure and meddle-free by design. Data can be distributed, but not copied. When it comes to digital assets and transactions, you can put almost anything on a Blockchain. Different scenarios call for different Blockchains. Figure 7.3 demonstrate how BC work [8]. Blockchain in government operates on the same principles as any blockchain application.

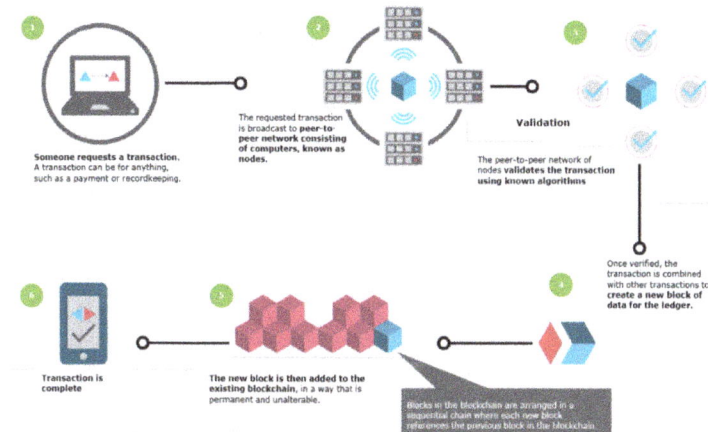

Figure 7.3 How a Blockchain works [8].

The BC technology currently has the following features [9,10]:

1. *Peer-to-Peer (P2P) Network*: The first requirement of BC is a network, an infrastructure shared by multiple parties. This can be a LAN at a small scale or the Internet at a large scale. All nodes participating in a BC are connected in a decentralized P2P network. Transactions are broadcast to the P2P network. Due to some limitations of P2P networks, some vendors have provided cloud-based BCs.

2. *Cascaded Encryption*: A BC uses encryption to protect transaction data. Blocks are encrypted in a cascaded manner, i.e. the encryption result of the previous block is used in encrypting the current block. The BC is secured by public

key cryptography, with each peer generating its own public-private key pairs.

3. *Distributed Database*: A BC is digitally distributed across a number of computers. Each party on a BC has access to the entire database and no single party controls the data or the information. Since BC is decentralized, there is no need for central authorizes such as banks.

4. *Transparency with Pseudonymity*: Each node or participant on a blockchain has a unique 30-plus-character alphanumeric address that identifies it. Users can choose to remain anonymous or provide proof of their identity to others.

5. *Irreversibility of Records*: Once a transaction is entered in the database and the accounts are updated, the records cannot be altered. Records on the database is permanent, chronologically ordered, and available to all others on the network.

As illustrated in Figure 7.4, a blockchain comprises a peer-to-peer network of participant nodes, a distributed ledger consisting of immutable blocks of data, transactions recorded in the blocks, smart contracts to execute the transactions, and a consensus algorithm that decides the proposer of the next block [11].

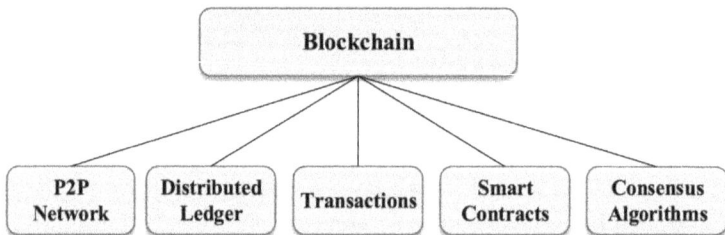

Figure 7.4 Components of a blockchain [11].

There are two types of blockchains: public and private. Public Blockchains are cryptocurrencies such as Bitcoin, enabling peer-to-peer transactions. Private Blockchains use Blockchain-based platforms such as Ethereum or Blockchain-as-a-service (BaaS) platforms running on private cloud infrastructure. A private BC is an intranet, while a public BC is the Internet. Companies will be disrupted the most by public Blockchains.

7.3 BLOCKCHAIN FOR GOVERNMENT

Blockchain refers to a range of general purpose technologies to exchange information and transact digital assets in distributed networks. It is regarded as one of the most important technology trends that will influence business and society in the years to come. The whole point of using a blockchain is to let people, especially those who do not trust themselves, share valuable data in a secure, tamperproof way. Blockchain technology has the potential to provide benefits to government as it enables reduced costs and complexity, shared trusted processes, improved discoverability of audit trials, and ensured trusted recordkeeping [12]. The development of blockchain technology is global. The blockchain revolution is taking place at a time when people are losing faith in governments and this new decentralized technology seems to be giving people new hope. There are a number of nations that have already started adopting blockchain technologies. Figure 7.5 shows Blockchain initiatives in several nations [13].

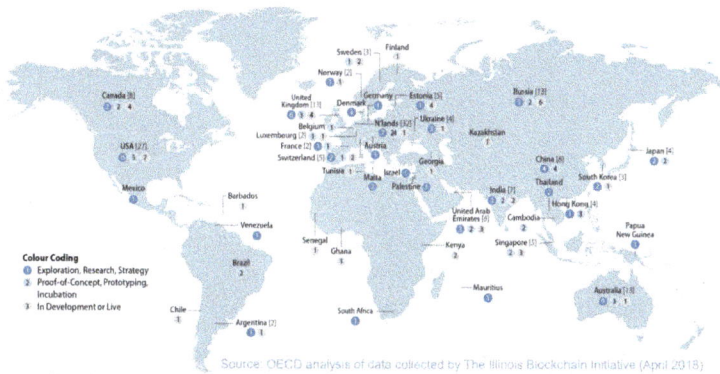

Figure 7.5 Blockchain initiatives in several nations [13]

Despite its proven real-world use cases, most government agencies continue to take a cautious approach to adopting blockchain into their paper-based processes. Examples of blockchain applications by governments include the following [14-16]:

- *Australia*: This nation teamed up with IBM as one of the top countries in the world integrating blockchain into its record keeping process. The country is using the technology to combine its passport and birth certificate databases into one blockchain-backed base.

• *China*: Despite initially taking an aggressive stance against cryptocurrencies, China has been quite open and positive about blockchain development. China recognizes that the technology can potentially usher in massive developments within various sectors in which it can be applied. It has announced its plan to build a "Blockchain city," based on blockchain technology. In China, the government has established a digital yuan, which will be used to replace some paper cash in circulation.

• *Denmark*: The government of Denmark is exploring blockchain technology to effectively collect taxes on used car sales. It has geared efforts towards developing blockchain-based voting platforms.

• *Estonia*: Estonia is the most digital country in the world, with 99% of public services are in the blockchain. It represents an ideal example of how the blockchain can benefit governments. Estonia has applied blockchain in its e-government system to ensure the safety and security of digital data and take a big step towards a truly digital state. The government of Estonia was one of the first to become a "blockchain government" by venturing into the technology. This has helped to significantly reduce the time that it takes for the government to deliver services to the country's citizens.

• *Ghana*: The Government of Ghana is using blockchain technology to solve two structural problems: determining the legal status of land ownership and solving longstanding miscommunication problems between the Ghanaian Land Commission, property owners, and financial institutions.

• *India*: India is one of the fastest-growing blockchain markets globally, with over 56% of Indian businesses reporting an inclination toward adopting blockchain technology. The country has plans to carry out research on the proactive use of blockchain technology. Due to the high need for safe business, the number of Blockchain companies in India is growing. The driving force behind India's blockchain adoption is the multitude of benefits it offers to enterprises and institutions.

• *Nigeria*: The Nigerian Federal Government has approved a national blockchain policy aimed at institutionalizing the technology in the country's economy and security sectors.

The approval of the national blockchain policy is a significant milestone, as it would ensure that blockchain technology is firmly embedded in Nigeria's economy. the Central Bank of Nigeria has used Blockchain to power the country's digital currency, eNaira. On 3 May 2023, the Nigerian government approved a National Blockchain Policy for Nigeria [17].

- *United Arab Emirates*: This is one of the countries that have fully embraced new technologies especially those that have a lot of potential for the future. The UAE has utilized blockchain technology for developing end-user services, relevant to the public sector. The government has launched a number of initiatives to promote the use of blockchain in the country. One of the most successful initiatives was the UAE Blockchain Strategy 2021, which helped the UAE government adopt the applications of blockchain in government and public sector to carry out its transactions.

- *South Korea*: The Korean government announced a 4B Korean won (about $3.5 million) award to set up a blockchain-enabled virtual power plant in the city of Busan, the country's second-most populous city. It has adopted blockchain for identity management. It is also actively developing the ICON platform, which is the country's largest blockchain project. The project aims to become a bridge between the online community and businesses in banking, healthcare, government, and many other industries.

- *United Kingdom*: The nation has shown interest in decentralized ledger technology, particularly with regards to the potential blockchain applications in the public sector. The British government has recognized that this technology can transform the conduct of public and private sector organizations. The new digital version of sterling, which has been given the unofficial name of "Britcoin," would give businesses and consumers the option to hold accounts directly with the Bank of England. Regulations in the United Kingdom allow residents to buy and sell cryptocurrencies.

- *United States*: In the US, blockchain adoption is happening at both the Federal level and also at the local level. DARPA and the Pentagon are reportedly working on coming up with strong security protocols using blockchain technology. Some policymakers seem to have dismissed blockchain. In March, 2023 the Economic Report of the President argued against any positive use of cryptocurrencies and the technologies behind them.

• *Venezuela*: This is the first federal government to introduce a digital currency in 2018: the petro. The petro has since alleviated the country's economic woes. It is acquired through the website of the treasury of cryptoactives of Venezuela. This currency is supported by oil, gold, diamonds, among others. The primary goal of the government is to eradicate the American currency (dollar), and ensure that citizens do not use it as a method of payment. Many Venezuelans are relying on cryptocurrencies to hedge against fiat inflation.

These are just a few examples of how blockchain technology is being used to improve the efficiency of government services. There are more use cases of blockchain being explored by governments around the world.

7.4 APPLICATIONS

Blockchain technology's key features, such as decentralization, transparency, immutability, and security, make it an attractive option for government applications. Blockchain can be used to improve the delivery of public sector services such as the issuance of title deeds, birth certificates, property transfers, business licensing, marital licenses, etc. Some of the uses of blockchain in government are shown in Figure 7.6 [14].

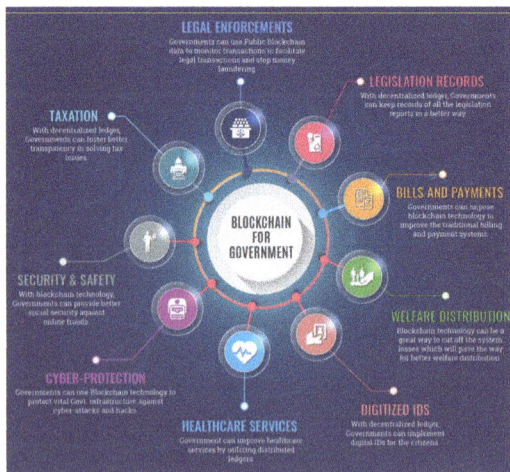

Figure 7.6   Uses of blockchain in government [14].

They are described as follows [14,18]:

- *Legal Enforcement*: Governments can use data from public to monitor money transactions and ensure that the system does not facilitate illegal dealings. The implementation of blockchain for a government can be a key tool towards making sure that financial transactions in the digital domain remain legal. This can be handy towards curbing illegal dealings such as drug trafficking and money laundering.

- *Taxation*: Blockchain might be the long-awaited solution for eliminating double taxation.

The decentralized ledger technologies have the capacity to foster more transparency in financial transactions. Governments can use decentralized ledger technologies to achieve more efficiency in its tax management systems through the integration of systems to track finances and eliminate double spending.

- *Government Infrastructure*: Blockchain can help governments to ensure better protection over their critical infrastructure, thereby keeping cyber attacks at bay. A decentralized ledger can also be developed such that it is able to keep track of the integrity of government systems. This would significantly reduce the chances of attacks and data tampering.

- *Welfare Distribution*: Decentralized ledger technology allows governments to handle matters related to welfare with more efficiency. Blockchain can be used to deploy faster service delivery allowing citizens to benefit directly. Such systems would also be more efficient in getting more people out of poverty. Blockchain technology can be used to facilitate proper allocation of funds to government projects.

- *Energy*: The US Department of Homeland Security defines 16 critical infrastructures including energy, food and agriculture, transportation, etc. Energy management can be daunting to governments. Governments can circumvent some of the issues revolving by coming up with efficient energy management systems that are based on the blockchain.

- *Tourism*: This is a very important sector for many nations. The US state of Hawaii has been looking for ways to develop blockchain solutions that can make it easier for tourists to enjoy what they have to offer. For example, cryptocurrencies can be used to make it easier for tourists to spend their money without having to worry about exchange rates.

- *Smart Cities*: A smart city uses information technology and data to integrate and manage physical, social, and business infrastructures to streamline services to its inhabitants. Blockchain technology can provide a safe and secure infrastructure for managing smart city applications such as energy management and transportation. For example, the city of Dubai is using blockchain to store and manage health records. By combining the security of blockchain with the efficiency of smart city technologies, governments are able to create a more seamless and secure experience for their citizens.

- *Central Banking*: Policy makers around the world are exploring ways to develop and issue central bank digital currencies in as little as the next five years. By eliminating the need for a central authority to issue and manage currency, blockchain could provide a more efficient, democratic alternative to traditional central banking.

- *Cross-Border Transactions*: Blockchain plays a crucial role at a place where security is needed the most. Border security is vital to every small and big nation. Globalization has made organizations to make more cross-border transactions. Blockchain has the potential to enable secure, efficient payments in cross-border transactions by removing the need for intermediaries. Multiple organizations are taking advantage of blockchain to enable cross-border transactions.

- *Cybersecurity*: Cybersecurity is ironically both a main benefit and potential weakness of blockchain technology. Data breaches do occur, and blockchain firms typically use these as learning lessons. Companies prefer to wait until the technology is more mature before entering.

- *Public Records*: Blockchain can be used to secure public records such as property titles, birth certificates, and business registrations. Once recorded on a blockchain, these records are immutable and cannot be altered or tampered with, ensuring their authenticity.

- *Regulatory Compliance*: Governments can use blockchain to monitor and enforce regulatory compliance in various industries. By recording relevant data on a blockchain, regulators can have real-time access to information, ensuring that businesses adhere to regulations.

- *Identity Management*: Governments can use blockchain for secure and efficient management of digital identities. Each citizen can have a unique digital identity stored on a blockchain, which can be used for accessing various government services.

- *Good Governance*: National governments are making efforts in de-globalizing various aspects of governance. The four major forms of governance include centralized, hybrid, polycentric, and decentralized. There is need for a polycentric governance model utilizing blockchain potentials where every participating member would abide by an agreed-upon set of rules. The issue of governance has been the focus of innovation. Blockchain governance might find a solution in a trusted execution environment (TEE). TEE can help data management across a decentralized system, where permissioned access to data is ensured through smart contracts. A key issue that governance needs to address is how collaborative innovation can be sustained between independent projects on the same network. This would require effective division of the right to control decisions among the core stakeholders. Figure 7.7 shows the components of good governance [19].

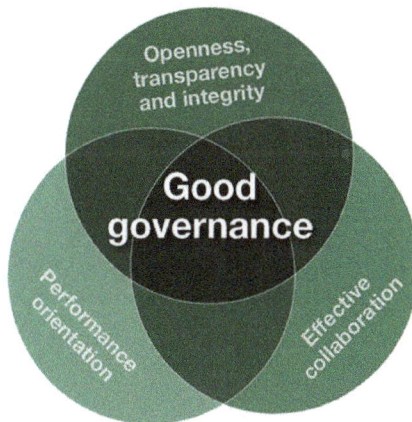

Figure 7.7   Components of good governance [19].

- *Anti-Corruption*:   Blockchain has the potential to be a game changer in anti-corruption efforts because it is designed to operate in environments where trust in data is greater than trust in individuals or institutions. Government corruption is the abuse of public power for private gain. It can assume various forms including bribery, embezzlement, and electoral fraud. At root, government corruption is a problem of trust. Corruption is commonly perceived as related to monetary benefits but it is much more in terms of misuse of power, coercion, disinformation, lack of transparency, nonperformance, inefficiency, delay tactics, and the lack of accountability. Governments regularly have to make trade-offs between efficiency and fairness in their services. Fair systems engender trust, pride, and a sense of community; while unfair systems encourage individuals to seek out illegal alternatives without remorse. In developing countries, blockchain applications for governments would come in handy towards eliminating. Blockchain can help dismantle corruption in government services. It could increase the fairness and efficiency of government systems while reducing opportunities for corruption [20].

• *Voting*: Voting remains an area that draws considerable negative attention due to corruption. Incidents of manipulated ballot boxes and voters disappearing from registries are common. The reliability of the voting process has been a problem since the beginning of democracy. Concerns about election protection, voter registration transparency, poll accessibility, and voter turnout have prompted governments to look at blockchain-based voting systems as a way to boost credibility and involvement in critical democratic processes. Implementing blockchain in voting systems can enhance the integrity of elections. Citizens can cast votes the same way they initiate other secure transactions and validate that their votes were cast, or even verify the election results. Blockchain technology can decentralize the voting process so that elections can happen securely with transparency. It is ideal for digital voting systems especially because transparency is one of its strong suites. Individuals who are eligible receive a token that allows them to vote only once and each vote is stored as a node in the blockchain.

• *Digital Currencies*: The digital currency will be used to conduct small transactions by consumers, companies, and authorities. While some governments have outrightly banned digital currencies, others are developing their own state-backed virtual currencies. Venezuela is in the lead, having launched its own state-backed cryptocurrency. Russia is also developing its state-backed cryptocurrency.

- *Smart Contracts*: These are programs that automatically fire when nodes come to consensus, without any human intervention. Smart contracts allow users to deploy new capabilities and functions while the blockchains are running; services do not have to be stopped. Specifically, developers could write a new smart contract that includes a set of function. The most popular smart contract platforms include the Ethereum Virtual Machine and Hyperledger Fabric's Chaincode. Smart contracts can improve efficiencies in contract enforcement.

Other areas of application include e-government, e-identities, education, personal identity security, public health, public procurement, public finance, business licensing, transportation, land registration and ownership, social benefits, citizen engagement, culture, humanitarian relief, social assistance, and grant disbursements.

7.5 BENEFITS

Blockchain is technology that builds a trustworthy service in an untrustworthy environment. Blockchain technology is already being used by governments around the world to improve the way they manage sensitive data. It is also safe against fraud and hacking so you do not have to worry about your information being stolen. Blockchain can be used to streamline administrative processes in the government by eliminating third parties such as agents. It can be adopted by governments as the technology through which they launch their pilot projects. It has been identified as the most suitable technology for digital migration. It is more affordable since there are no middlemen to introduce markups. Many companies and banking institutions have been adopting blockchain technology to achieve more efficiency. Other benefits of blockchain government include the following [21]:

- *Social Trust*: There are risks with any system. An important function of government is to maintain trusted information about individuals, organizations, assets, and activities. Social trust in the technology is an issue. Most people do not trust their government.

Blockchain, with all its different features, comes as the solution to this problem. It is technology that builds a trustworthy service in an untrustworthy environment. Transparency changes some of the sentiments by allowing citizens to view and verify data. Blockchain can help build trust in new vaccines from inception to injection with a network powered by IBM Blockchain, such as the vaccine system in Figure 7.8 [21].

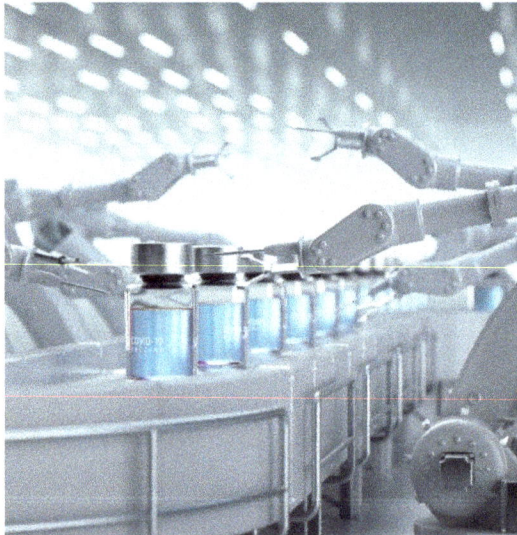

Figure 7.8   Blockchain can help build trust in new vaccines [21].

Through the use of blockchain, all the parties involved in a transaction only have to trust the blockchain without a need for a central intermediary.

• *Cost Reduction*: The use of blockchain can lower the costs of government services and also limit redundancy, streamline processes, increase security, decrease the audit burden and even ensure that the data integrity is maintained.

• *Social Evils*: Governments are concerned that blockchain technology and cryptocurrencies are facilitating social evils such as money laundering, funding criminal activities, corruption, and tax evasion

• *Eliminating Corruption*: Combatting corruption in public procurement is another area where blockchains have been implemented and show promise for expansion. The costs to society of public-sector corruption and weak accountability are staggering. Blockchain technology serves a unique role in combating government corruption. In developing countries, blockchain applications for governments can help eliminate the main issue of corruption while also ensuring more effective deployment and distribution of resources.

- *Fraud*: The blockchain is safe against fraud and hacking so you do not have to worry about your information being stolen. There is a lot of stored government records on the blockchain, including land titles, birth certificates, and marriage licenses. Election fraud continues to be a major issue for governments around the world. Voting machine hacks and intentional voting miscounts from groups in power.

7.6 CHALLENGES

Although blockchain technology holds great promise, its implementation in government systems also comes with challenges, such as scalability, interoperability, and regulatory issues. Blockchains cannot solve all problems. There is no one-size-fits-all blockchain system. Blockchain is not mature yet, as challenges exist for both adoption and technology development. It has yet to reach widespread adoption at scale. Other challenges include the following [22].

- *Security*: Defense and security are the major concerns of every nation. One of the major hurdles that have stood in the way of digitization for many governments is security. Bringing personal data of millions of people onto digital platforms presents a huge risk when the system is hacked. Several blockchain systems intentionally make their consensus protocols proprietary, making it difficult to trust in the correctness and security of the platforms.

- *Data quality*: Blockchain does not protect against data from untrustworthy sources. It cannot prevent well-formatted but incorrect or inaccurate data from being sent and stored in the system.

- *Regulations*: When it comes to implementing and adopting blockchain by government institutions, the major hurdle seems to be the lack of standards and regulations.

Regulating cryptocurrencies has proved to be a nightmare for governments.

- *Scalability*: This may be interpreted as the number of nodes and the number of clients. The number of nodes is a concern during blockchain deployment: how many nodes should one use to start the service? Both permissionless and permissioned blockchains have scalability limits.

- *Privacy and Compliance*: These are always major concerns in governmental applications. Although conventional blockchains provide availability and integrity, the data are essentially transparent—all participants may freely review transactions. At the same time, laws designed to safeguard the privacy and security of individuals' information do provide a roadmap for designers

7.7 CONCLUSION

Blockchain technology is no longer limited to digital currencies. Today, many real-world blockchain government use cases have emerged. The excitement around using blockchain in the public sector is building at a large scale around the world. Blockchain technology could accelerate key government functions such as voting, identity verification, certifying transactions such for land-use registry, and safekeeping of medical records. It is an excellent tool that governments worldwide can use to build systems that are open and trustworthy and protect the citizens of their country. It provides governments with a fast, secure, efficient, and transparent way to deliver government services and communicate with their citizens.

Blockchain technology has the potential to revolutionize the way governments operate and interact with citizens. Governments across the globe have taken different approaches to developing their blockchain ecosystems. Blockchains have been hailed as a building block for a new economy due to the security and decentralization that it provides. Deemed by experts as the most important technological innovation since the Internet, blockchain is set to revolutionize the global economy. The Government Blockchain Association (GBA) is an international nonprofit, professional membership association, connecting individuals and organizations with blockchain technology solutions to government requirements. The GBA is all about driving authentic connectivity that the global initiatives support. GBA members comprise of over 2,000 leading industry experts representing the most comprehensive blockchain technology capacities in the world [23]. Many governments hope that blockchain will be a game changer for issues such as security and operational challenges. More information on blockchain government can be found in the books in [24-36] and related journals:

- *Blockchain: Research and Applications.*
- *Digital Government: Research and Practice*

REFERENCES

[1] M. N. O. Sadiku, C. M. M. Kotteti, and J. O. Sadiku, "Blockchain in government: A primer," Journal of Scientific and Engineering Research, vol. 10, no. 9, 2023, pp. 1-10.

[2] S. Ølnes, J. Ubacht, and M. Janssen, "Blockchain in government: Benefits and implications of distributed ledger technology for information sharing," Government Information Quarterly: An International Journal of Information Technology Management, Policies, and Practices, vol. 34, no. 3, 2017, pp. 355-364.

[3] H. Bamakan et al., "Blockchain technology forecasting by patent analytics and text mining," Blockchain: Research and Applications, vol. 2, no. 2, June 2021.

[4] "Blockchain in government and the public sector,"

https://consensys.net/blockchain-use-cases/government-and-the-public-sector/

[5] M. N. O. Sadiku, Y. Wang, S. Cui, and S. M. Musa, "A primer on blockchain," International Journal of Advances in Scientific Research and Engineering, vol. 4, no. 2, February 2018, pp. 40-44.

[6] C. Singh, R. Thakkar, and J. Warraich, "Blockchain in supply chain management," European Journal of Engineering and Technology Research, vol. 7, no. 5, October 2022, pp. 60-70.

[7] S. Depolo, "Why you should care about blockchains: the non-financial uses of blockchain technology," March 2016,

https://www.nesta.org.uk/blog/why-you-should-care-about-blockchains-non-financial-uses-blockchain-technology

[8] J. Schmitz, "Blockchain technology for accounting and government,"

https://na.eventscloud.com/file_uploads/6c2267c4bafdc5487eac41e98af7c9e4_1530JanaSchmitz.pdf

[9] M. Iansiti and K. R. Lakhani, "The truth about blockchain," Harvard Business Review, Jan./Feb. 2017.

https://hbr.org/2017/01/the-truth-about-blockchain

[10] W. T. Tsai et al., "A system view of financial blockchains," Proceedings of IEEE Symposium on Service-Oriented System Engineering, 2016, pp. 450-457.

[11] D. Shakhbulatov et al., "How blockchain enhances supply chain management: A survey," IEEE Open Journal of the Computer Society, vol. 1, 2020, pp. 230-249.

[12] "Benefits of Telos blockchain in government processes," July 2019,

https://medium.com/@Telosfeed_en/benefits-of-telos-blockchain-in-government-processes-2e2f6d2cad4

[13] "Uses and limitations of blockchain in the public sector," October 2018,

https://www.oecd.org/parliamentarians/meetings/gpn-meeting-october-2018/OPSI-Blockchain-Presentation-for-Global-Parliamentary-Network.pdf

[14] H. Anwar, "Blockchain government transformation: What it means? And how it will improve our life?" August 2018,

https://101blockchains.com/blockchain-government-transformation/

[15] "What are the benefits of blockchain for government services?" July 2022,

https://appinventiv.com/blog/role-of-blockchain-in-government/

[16] P. Bustamante et al., "Government by code? Blockchain applications to public sector governance," Frontier in Blockchain, vol. 5, June 2022.

[17] O. Adesina, "Nigerian government approves the use of blockchain,"

https://nairametrics.com/2023/05/03/official-nigerian-government-approves-the-use-of-blockchain/

[18] "Top 17 blockchain applications & use cases in 2023," January 2023,

https://research.aimultiple.com/blockchain-applications/

[19] "Benefits of blockchain: A business sector perspective,"

https://www.e-zigurat.com/innovation-school/blog/benefits-of-blockchain/

[20] M. Niekerk, "How blockchain can help dismantle corruption in government services," August 2022,

https://www.thepeninsula.org.in/2022/08/27/how-blockchain-can-help-dismantle-corruption-in-government-services/?print=print

[21] "Blockchain for government,"

https://www.ibm.com/blockchain/industries/government

[22] J. Clavin et al., "Blockchains for government: Use cases and challenges," Digital Government: Research and Practice, vol. 1, no. 3, Article 22, November 2020.

[23] "Government Blockchain Association,"

https://www.linkedin.com/company/gbaglobal

[24] M. Jun, Blockchain Government: A Next Form of Infrastructure for The Twenty-First Century. CreateSpace Independent Publishing Platform, 2018.

[25] Information Resources Management Association, Research Anthology on Blockchain Technology in Business, Healthcare, Education, and Government. ⌈IGI Global, 2020.

[26] F. Pignatelli (ed.), Blockchain for Digital Government: An Assessment of Pioneering Implementations in Public Services. Luxembourg: Publications Office of the European Union, 2019.

[27] K. Saini and M. Khari (eds.), Handbook of Green Computing and Blockchain Technologies. Boca Raton, FL: CRC Press, 2021.

[28] M. Ghonge and N. Pradeep, Blockchain Technologies and Applications for Digital Governance. IGI Global, 2021.

[29] C. G. Reddick, H. J. Scholl, and M. P. Rodríguez-Bolívar, Blockchain and the Public Sector: Theories, Reforms, and Case Studies. Springer, 2021.

[30] M. Mathiesen, The Blockchain Government: Engineering the Future. Independently Published, 2021.

[31] I. Williams, Cross-Industry Use of Blockchain Technology and Opportunities for the Future. IGI Global, 2020.

[32] G. Blokdyk, Blockchain in Government a Complete

Guide. Emereo Pty Limited, 2018.

[33] S. Mahankali and R. Dhuddu, Blockchain in E-Governance: Driving the Next Frontier in G2C Services. BPB PUBN, 2021.

[34] J. Lindman, The Uncertain Promise of Blockchain for Government. OECD Publishing, 2020.

[35] C. Babaoğlu, E. Akman, and O. Kulaç, Handbook of Research on Global Challenges for Improving Public Services and Government Operations. Information Science Reference, 2020.

[36] M. Jun, Blockchain Government: A Next Form of Infrastructure for the Twenty-First Century. CreateSpace Independent Publishing Platform, 2018.

CHAPTER 8

BLOCKCHAIN IN INTERNET OF THINGS

*"The Blockchain cannot be described just as a revolution. It is
a tsunami-like phenomenon, slowly advancing and gradually
enveloping everything along its way by the force of its progression."*

— William Mougaya

8.1 INTRODUCTION

Advances in technology always have an impact on our society. Emerging technologies, such as the Internet of Things (IoT) and Blockchain, present transformative opportunities. A Blockchain is a decentralized network consisting of data records or "blocks" that cannot be modified by a single actor. How the information is stored is difficult or nearly impossible to hack, change, or cheat the system. That is because Blockchain offers a supremely robust level of encryption. Blockchain is used in many areas as illustrated in Figure 8.1 [1].

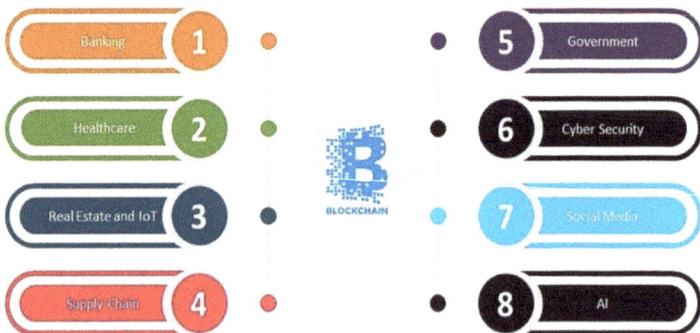

Figure 8.1   Blockchain is used in many areas [1].

It is a technology that is currently receiving great attention and may help in providing security in IoT scenarios. Unlike the Internet alone, Blockchains are distributed, not centralized; open, not hidden; inclusive, not exclusive; immutable, not alterable; and secure [2].

The world is becoming more and more connected, but increased vulnerability comes with increased connectivity. The Internet of things (IoT) connects people, places, and products, offering new opportunities to generate value in products and business processes. IoT is a network of connected devices and people - collecting and sharing data. It creates a web of interconnected devices that are exposed to hackers and vulnerable to attack. This network of connected devices is transforming the way enterprises operate through the use of sensors and other edge devices and infrastructure. It has the potential to revolutionize the way we live and work by automating mundane tasks. IoT holds the promise of connecting almost any device to the Internet and making the devices smarter and more accessible. However, IoT devices are prone to cyber attacks due to their lack of built-in security measures. Security is a major concern when implementing large-scale deployment of IoT. Blockchain technology could play a crucial role in enhancing the security of the IoT [3]. Due to the decentralized nature of Blockchain, there is no central point of control, making it almost impossible to hack or tamper with. The combination of IoT with Blockchain refers to the use of a cryptographically secure digital ledger to authenticate, store, and share data generated by connected IoT devices. Integrating Blockchain with the Internet of things (IoT) can deliver numerous benefits, such as improving transparency, traceability, reliability, and automation. However, combining the two technologies also comes with challenges, including scalability, interoperability, and energy consumption [4].

This chapter provides a short review of the union of IoT and Blockchain. It begins with presenting an overview on Blockchain and IoT to make the chapter self-contained. It considers the integration of Blockchain with IoT. It covers some applications of BC in IoT. It highlights the benefits and challenges of BC in IoT. The last section concludes with comments.

8.2 WHAT IS BLOCKCHAIN?

Blockchain (BC) technology is a permanent record of online transactions. It is a distributed tamper-proof database, shared, and maintained by multiple parties. It is a new enabling technology that is expected to revolutionize many industries, including business. It has the potential for addressing significant business issues. The BC technology allows participants to move data in real-time, without exposing the channels to theft, forgery, and malice.

The term "Blockchain" refers to the way BC stores transaction data – in "blocks" that are linked together to form a "chain." The chain grows as the number of transactions increases. Since every entry is stored as a block on a chain, the care you receive is added to your personal ledger. The first Blockchain was conceived in 2008 by an anonymous person or group known as Satoshi Nakamoto, who published a white paper introducing the concept of a peer-to-peer electronic cash system he called Bitcoin [5]. Blockchain is a distributed ledger database that consists of records or transactions or various digital incidents that are executed by the participants. Blockchain technology can assist in achieving the seven objectives of SCM: their cost, quality, speed, dependency, risk reduction, sustainability as well as flexibility. The concept of Blockchain is shown in Figure 8.2 [6]. Figure 8.3 shows how Blockchain works [7].

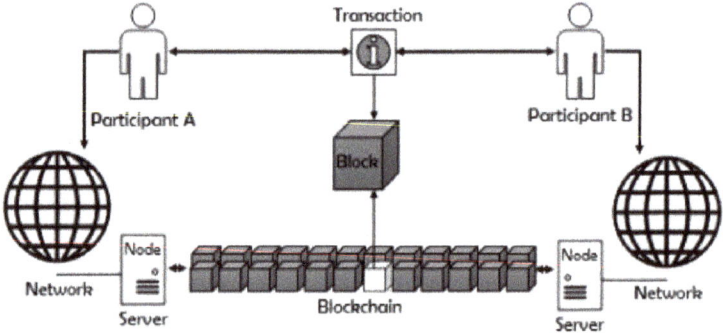

Figure 8.2 The concept of Blockchain [6].

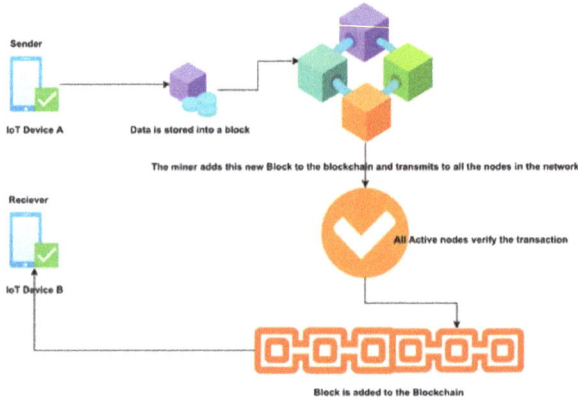

Figure 8.3 How Blockchain works [7].

At its core, Blockchain is a distributed system recording and storing transaction records. In a Blockchain system, there is no central authority. Instead, transaction records are stored and distributed across all network participants. Rather than having a centrally located database that manages records, the database is distributed to the networks and transactions are kept secure via cryptography. BC eliminates the need for a middleman that traditionally may facilitate such transactions.

Fundamentally, Blockchains are distributed digital database that record and maintain a list of transactions taking place in real time. They may also be regarded as decentralized ledgers that sequentially record transactions or interactions among users within a distributed network. They have the following properties [8]:

- Firstly, they are autonomous. They run on their own, without any person or company in charge.

- Secondly, they are permanent. They are like global computers with 100 percent uptime. Because the contents of the database are copied across thousands of computers, if 99 per cent of the computers running it were taken offline, the records would remain accessible and the network could rebuild itself.

- Thirdly, they are secure and tamper-proof. Each record in Blockchain is time stamped and stored cryptographically. The encryption used on Blockchains like Bitcoin and Ethereum is industry standard, open source, and has never been broken.

- Fourthly, they are open, allowing anyone to develop products and services on them.

- Fifthly, as Blockchain is a shared system, costs are also shared between all of its users.

Some characteristics of Blockchain are portrayed in Figure 8.4 [9].

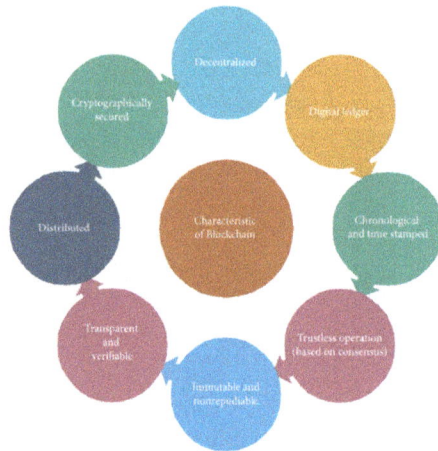

Figure 8.4   Some characteristics of Blockchain [9].

The Blockchain was designed so transactions are immutable, i.e. they cannot be deleted. Thus, Blockchains are secure and meddle-free by design. Data can be distributed, but not copied. When it comes to digital assets and transactions, you can put almost anything on a Blockchain. Different scenarios call for different Blockchains [10].

The BC technology currently has the following features [11-13]:

1. *Peer-to-Peer (P2P) Network*: The first requirement of BC is a network, an infrastructure shared by multiple parties. This can be a LAN at a small scale or the Internet at a large scale. All nodes participating in a BC are connected in a decentralized P2P network. Transactions are broadcast to the P2P network. Due to some limitations of P2P networks, some vendors have provided cloud-based BCs.

2. *Cascaded Encryption*: A BC uses encryption to protect transaction data. Blocks are encrypted in a cascaded manner, i.e. the encryption result of the previous block is used in encrypting the current block. The BC is secured by public key cryptography, with each peer generating its own public-private key pairs.

3. *Distributed Database*: A BC is digitally distributed across a number of computers. Each party on a BC has access to the

entire database and no single party controls the data or the information. Since BC is decentralized, there is no need for central authorizes such as banks.

4. *Transparency with Pseudonymity*: Each node or participant on a Blockchain has a unique 30-plus-character alphanumeric address that identifies it. Users can choose to remain anonymous or provide proof of their identity to others.

5. *Irreversibility of Records*: Once a transaction is entered in the database and the accounts are updated, the records cannot be altered. Records on the database is permanent, chronologically ordered, and available to all others on the network.

A Blockchain comprises a peer-to-peer network of participant nodes, a distributed ledger consisting of immutable blocks of data, transactions recorded in the blocks, smart contracts to execute the transactions, and a consensus algorithm that decides the proposer of the next block [10].There are two types of Blockchains: public and private. Public Blockchains are cryptocurrencies such as Bitcoin, enabling peer-to-peer transactions. Private Blockchains use Blockchain-based platforms such as Ethereum or Blockchain-as-a-service (BaaS) platforms running on private cloud infrastructure. A private BC is an intranet, while a public BC is the Internet. Companies will be disrupted the most by public Blockchains.

8.3 OVERVIEW ON INTERNET OF THINGS
The term "Internet of things" was introduced by Kevin Ashton from the United Kingdom in 1999. Internet of Things (IoT) is a network of connecting devices embedded with sensors. It is a collection of identifiable things with the ability to communicate over wired or wireless networks.   It is the global interconnection of several heterogeneous devices. The devices or things can be connected to the Internet through three main technology components: physical devices and sensors (connected things), connection and infrastructure, and analytics and applications. The Internet of things can be defined using seven characteristics in Figure 8.5 [14].

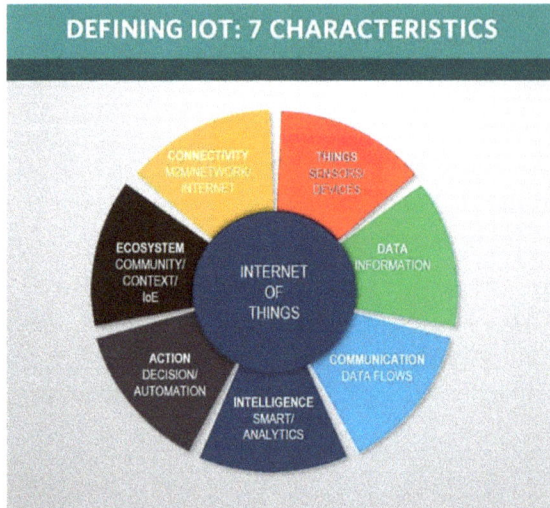

Figure 8.5   The Internet of things is defined using 7 characteristics [14].

The IoT, also known as the Internet of Objects, or the Internet of everything, or the Web of Objects, is a worldwide network that connects devices to the Internet and to each other using wireless technology. It has been gaining popularity rapidly since its inception into the IT community and is being used in healthcare, education, gaming, finance, transportation, and several more. IoT is expanding rapidly and it has been estimated that 50 billion devices will be connected to the Internet by 2020. These include smart phones, tablets, desktop computers, autonomous vehicles, refrigerators, toasters, thermostats, cameras, pet monitors, alarm systems, home appliances, insulin pumps, industrial machines, intelligent wheelchairs, wireless sensors, mobile robots, etc.

There are four main technologies that enable IoT [15]:

(1) Radio-frequency identification (RFID) and near-field communication.

(2) Optical tags and quick response codes.

(3) Bluetooth low energy (BLE).

(4) Wireless sensor network.

Other related technologies are cloud computing, machine learning, and big data.

The concept of IoT has some the following characteristics [16]:

- *Interconnected*: Internet of things facilitates people to devices and devices to other devices.

- *Smart sensing*: The majority of devices and actuators have embedded or connected sensors to detect current conditions.

- *Intelligence*: IoT devices have some calculating units and software used for smart decisions, predictions and automation control.

- *Energetical Efficiency*: All IoT devices must be efficient and able to use recyclable energy, boost own energy harvesting, if the application of device requires and allows it.

- *Data Sharing*: IoT connected devices have the capability to express and share their current state to all other connected devices.

- *Safety*: Internet of things devices should ensure the safety of individual life. All medical smart devices are a good example of this characteristic.

IoT supports many input-output devices such as camera, microphone, keyboard, speaker, displays, microcontrollers, and transceivers. It is the most promising trend in the healthcare industry. Today, smartphone acts as the main driver of IoT.

The narrowband version of IoT is known as narrowband IoT (NBIoT). This is an attractive technology for many sectors including healthcare because it has been standardized [17]. The main feature of NBIoT is that it can be easily deployed within the current cellular infrastructure with a software upgrade.

8.4 BLOCKCHAIN FOR IOT

Both Blockchain and IoT are emerging technologies with great potential, but still lacking widespread adoption due to technical and security concerns. IoT and Blockchain work hand in hand to solve several problems. The current centralized model of IoT deployments has many security challenges. One way to solve this is to decentralize the IoT network by establishing a permission-based ledger.

Blockchain technology is a game-changer in the realm of data security and has the potential to revolutionize the way we interact with our connected devices. It can provide enhanced data privacy and security to IoT devices. IoT enables connected devices across the Internet to transmit data to Blockchain networks and create tamper-resistant records of transactions in the process. IoT enhances security and transparency in IoT ecosystems. By integrating IoT with Blockchain technology, IoT data can be secured, authenticated, and decentralized, which improves trust, transparency, traceability, and reliability in IoT-based processes and automation. The combination of IoT and Blockchain allows a smart device to function autonomously. It can protect organizations and individuals from different forms of harm. It can improve the reliability and traceability of the network. Figure 8.6 displays some Blockchain-IoT opportunities [18].

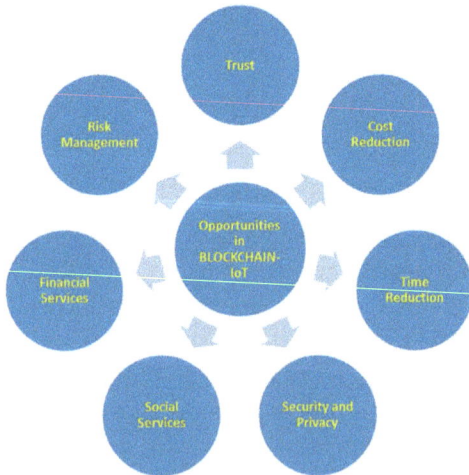

Figure 8.6   Blockchain-IoT opportunities [18].

8.5 APPLICATIONS OF BLOCKCHAIN IN IOT

Blockchain technology can be used to secure IoT systems in various ways. Several industries have started to explore the potential applications of IoT and Blockchain to improve efficiency and bring automation. Some examples presented as follows [19,20]:

- *Banking*: Several banks are introducing Blockchain technology to create a scalable and decentralized environment for IoT devices and applications. Banks are ameliorating their services by utilizing new Blockchain protocols. Today, banks can transact directly with each other at low costs of settlements.

- *Supply Chain*: The ability to track components used in aircraft, automobiles, and other products is essential for enabling regulatory compliance and safety. Many companies are working to enable IoT devices to track the shipping process. By integrating IoT devices such as sensors and RFID tags with a Blockchain network, companies can create a decentralized and secure system for tracking products as they move through the supply chain. Due to the lack of transparency, the combination of both IoT and Blockchain technologies improves the reliability and traceability of the network.

- *Smart Cities*: These are also harnessing the power of Blockchain-enabled IoT systems. As more cities begin to implement IoT strategy within their infrastructure, Blockchain has the potential to enhance use cases. In the future, IoT can potentially undergird the infrastructure of smart cities in order to make communications far more streamlined and efficient than they are today. The union of IoT and Blockchain technology makes it possible to remotely manage your security system, eliminate centralized infrastructure.

- *Smart Homes*: Smart IoT-enabled devices play a crucial role in our daily life. Blockchain and IoT are the future of smart home technology. In a smart home, the devices are connected and can be accessed through the central point such as a gaming console, smartphone, tablet, or laptop. To enable these smart homes, the role of IoT is crucial. Integrating IoT and Blockchain technology makes it possible to remotely manage your security system. For example, Telstra, an Australian telecommunication and media company, provides smart home solutions. The company has implemented Blockchain to ensure no one can manipulate the data captured from smart devices.

- *Automotive Industry*: This industry is investing in developing automated vehicles enabled by IoT sensors, allowing crucial information to be exchanged easily and quickly. It is using IoT-enabled sensors to develop fully automated vehicles. Combined with Blockchain, autonomous cars, smart parking, or automated traffic control can be achieved.

- *Pharmaceutical Industry*: The problem of counterfeit products in the pharma industry is increasing daily. The transparency and traceability of Blockchain technology can help to control the shipment of medicines from their origin to their point of destination.

- *Agriculture*: Blockchain technology has the potential to reshape agriculture, from production to the grocery store. IoT sensors can be installed on farms and send their data directly to a Blockchain network to improve the supply chain. Blockchain for supply chain management is transforming the agricultural industry by improving food safety and quality, as well as traceability of the entire agricultural supply chain.

- *Freight Transportation*: Freight is a complex process involving different parties with different priorities. Moving freight from one place to another is complicated. With the help of an IoT-enabled Blockchain, it is possible to store temperatures, arrival times, and the status of shipping containers in transit. Immutable Blockchain transactions help ensure that everyone involved in the process can trust the data and take action to move products quickly and efficiently.

- *Cybersecurity*: Security is one of the most important issues with the IoT and it is an ongoing problem. It will continue to evolve as regulations related to their development and use continue to match forward. However, the possibility of a Blockchain IoT security system is something that may hold great potential. Security is one of the innate attributes of Blockchain and the technology has the ability to legitimize data and make sure it comes from a trusted source. Blockchain can be used to safeguard IoT devices against cyberattacks by providing various security measures.

- *Privacy*: In IoT, security and privacy remain a major challenge due to the massive scale and distributed nature of IoT networks. IoT devices integrated with Blockchain can improve privacy.

- *Automated Transactions*: IoT devices can be programmed to automatically trigger transactions on the Blockchain, leading to more efficient and streamlined processes.

These are just a few examples, but the potential of the combination of IoT and Blockchain goes far beyond these applications. Figure 8.7 shows some applications of Blockchain in IoT [21].

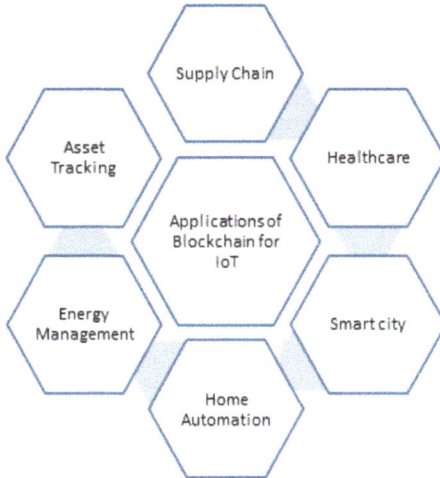

Figure 8.7   Some applications of Blockchain in IoT [21].

8.6 BENEFITS

One of the most attractive characteristics of Blockchain is its ability to secure data and thwart cyber attacks. Blockchain technology can help ease the associated security and scalability issues. It brings transparency by allowing any authorized person to access the network and track past transactions. Other benefits of combining Blockchain with IoT are [20]:

- *Data Authenticity for Quality Assurance*: Blockchain technology can add a robust framework to processes to quickly and accurately detect data manipulations.

- *Device Tracking to Catch Errors*: IoT networks can be gigantic, making failure patterns difficult to detect. Blockchain technology assigns a unique key to each IoT endpoint that helps identify inconsistencies.

- *Usage Logs for Employee Performance*: Blockchain technology can track user actions to let you see who, when, and how employees have used a device.

- *Consumer Transparency*: Many consumers are now in

the dark when it comes to knowing where their data is kept, who can access it, or how it's transferred. Blockchain would equip them with this knowledge, since it saves and logs all communications that happen within the user's IoT devices.

- *Efficiency*: IoT devices automate and streamline routine tasks, leading to significant improvements in efficiency and productivity.

- *Real-time Monitoring*: An IoT ecosystem collects and analyzes data in real time, providing valuable insights that can aid quick informed decision-making.

- *Enhanced Security*: The Internet of Things can also contribute to safety and security. Blockchain's decentralized nature and use of cryptography significantly bolsters IoT security. Traceability and validation are enforced by multiple nodes in the network, making it extremely difficult for hackers and cybercriminals to alter or falsify data.

- *Decentralization*: Unlike traditional IoT models that rely on a central server, Blockchains operate on a peer-to-peer network. This decentralization eliminates single points of failure and can improve the scalability and efficiency of IoT systems.

- *Trust*: One of the main benefits of Blockchain is its immutable and transparent nature, which helps to foster trust among users. Every transaction is recorded and can be traced back, providing a reliable audit trail.

- *Smart Contracts*: IoT technology alone enables automation. When coupled with smart contracts, fast automated responses can be authorized through this network. Blockchain enables the use of smart contracts with the terms of the agreement directly written into code. Smart contracts can automate various processes in an IoT network, improving efficiency and reducing the need for intermediaries.

Some of these benefits are shown in Figure 8.8 [20].

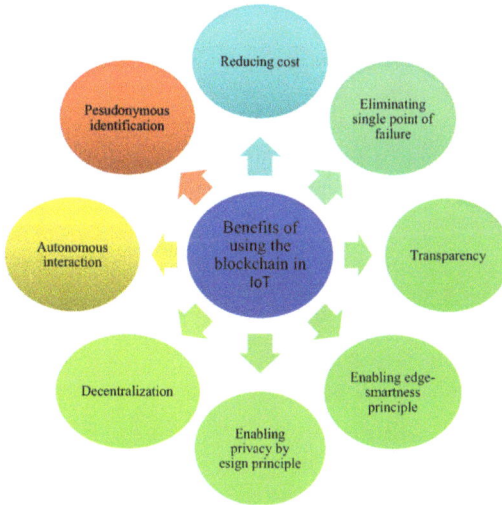

Figure 8.8   Benefits of Blockchain in IoT [20].

8.7 CHALLENGES

Adopting IoT and Blockchain technologies is not as widespread as one would believe. That is due to operational challenges and technical concerns. Security is a major concern with IoT that has hindered its large-scale deployment. Blockchain technology   has the potential to help address some of the IoT security and scalability challenges. Other challenges include the following [17, 18]:

- *Hacking*: Cybercriminals always look for known vulnerabilities in IoT devices. As organizations are getting more dependent on the Internet, the scope of hackers to exploit businesses has risen exponentially. With millions of devices connected by IoT, the technology is an open target for hacking. The Mirai botnet attack of 2016 is a prime example, using thousands of compromised household IoT devices to launch a devastating distributed denial of service (DDoS) campaign against high-profile sites and services. Blockchain technology can help by making IoT devices more secure and efficient.

- *Weak Passwords*: Using weak and hard-coded passwords in IoT devices is a major security challenge facing the industry. These types of passwords can be easily cracked by attackers, granting them access to the devices. Managing passwords in a distributed IoT ecosystem is difficult, especially when devices are managed remotely.

- *Insecure Network*: Hackers target IoT devices by exploiting weaknesses in communication protocols and services to access sensitive information. Man-in-the-middle is an example of an attack resulting from an insecure network. It is crucial to secure IoT communications using best practices in the industry.

- *Lack of Standardization in IoT*: The absence of standardization in IoT security is a major concern. With so many different devices from various manufacturers, it can take time to ensure that every single device has the proper level of security. This makes it harder to implement security measures and allows hackers to exploit vulnerabilities. Security professionals may not be familiar with the specific protocols and technologies used by different devices.

- *Scalability*: Another challenge with current IoT networks is that of scalability, how to handle the massive amounts of data collected by a large network of sensors. As the number of IoT devices grows, ensuring that Blockchain networks can effectively handle the increasing volume of transactions is a barrier.

These challenges are displayed in Figure 8.9 [22].

Figure 8.9 Some challenges of Blockchain in IoT [22].

8.8 CONCLUSION

IoT devices often suffer with security vulnerabilities that make them an easy target for DDoS attacks. The union of IoT and Blockchain technology is transforming the way we communicate and interact. The merging of Blockchain and the IoT is also expected to address the issue security and privacy.

Chain of things (CoT) is a consortium of technologists and leading Blockchain companies. It investigates the best possible use cases where a combination of Blockchain and IoT can offer significant benefits to industrial, environmental, and humanitarian applications. More information on Blockchain in IoT can be found in the books in [23-47] and the following related journal:

- *IEEE Internet of Things Journal*
- *Blockchain: Research and Applications*

REFERENCES

[1] R. Raj, "Top Blockchain applications - A comprehensive guide," August 2023,

 https://intellipaat.com/blog/tutorial/Blockchain-tutorial/Blockchain-applications/

[2] "IoT powered by Blockchain: How Blockchains facilitate the application of digital twins in IoT,"

https://www2.deloitte.com/content/dam/Deloitte/de/Documents/Innovation/IoT-powered-by-Blockchain-Deloitte.pdf

[3] A. Dorri et al., "Blockchain for IoT security and privacy: The case study of a smart home,"

Proceedings of 2ND IEEE PERCOM Workshop On Security Privacy And Trust In The Internet of Things, 2017.

[4] M. N. O. Sadiku, U. C. Chukwu, and J. O. Sadiku, "Blockchain in IoT: A Brief Review," Web of Scholar: Multidimensional Research Journal, vol. 2, no. 8, 2023, pp. 97-107.

[5] M. N. O. Sadiku, Y. Wang, S. Cui, and S. M. Musa, "A primer on Blockchain," International Journal of Advances in Scientific Research and Engineering, vol. 4, no. 2, February 2018, pp. 40-44.

[6] C. Singh, R. Thakkar, and J. Warraich, "Blockchain in supply chain management," European Journal of Engineering and Technology Research, vol. 7,no. 5, October 2022, pp. 60-70.

[7] T. Alam, "Blockchain-based internet of things: Review, current trends, applications, and future challenges," Computers, vol. 12, no.1, 2023.

[8] S. Depolo, "Why you should care about Blockchains: the non-financial uses of Blockchain technology," March 2016,

https://www.nesta.org.uk/blog/why-you-should-care-about-Blockchains-non-financial-uses-Blockchain-technology

[9] P. Ratta et al., "Application of Blockchain and Internet of things in healthcare and medical sector: applications, challenges, and future perspectives," Journal of Food Quality, vol. 2021, May 2021.

[10] "Benefits of Blockchain: A business sector perspective,"
https://www.e-zigurat.com/innovation-school/blog/benefits-of-Blockchain/

[11] M. Iansiti and K. R. Lakhani, "The truth about Blockchain," Harvard Business Review, Jan./Feb. 2017.
https://hbr.org/2017/01/the-truth-about-Blockchain

[12] W. T. Tsai et al., "A system view of financial Blockchains," Proceedings of IEEE Symposium on Service-Oriented System Engineering, 2016, pp. 450-457.

[13] D. Shakhbulatov et al., "How Blockchain enhances supply chain management: A survey," IEEE Open Journal of the Computer Society, vol. 1, 2020, pp. 230-249.

[14] S. Merenych, "IoT in business in 2023: Benefits of implementing Internet of things," January 2023,
https://clockwise.software/blog/iot-in-business-benefits-of-internet-of-things/

[15] "Top 15 Benefits of IoT in business [updated 2021]," March 2021,
https://www.mobinius.com/blogs/iot-benefits-in-bussiness-2021/

[16] Q. F. Hasan, A. R. Khan, and S. A. Madani (eds.), Internet of Things: Challenges, Advances, and Applications. Boca Raton, FL: CRC Press, 2018.

[17] N. M.Kumara and P. K. Mallickb, "Blockchain technology for security issues and challenges in IoT," Procedia Computer Science, vol. 132, 2018, pp. 1815–1823.

[18] O. Nwosu, "The key to security: Combining IOT and Blockchain technology," February 2023,
https://www.analyticsvidhya.com/blog/2023/02/the-key-to-security-combining-iot-and-Blockchain-technology/

[19] E. Canorea, "How is Blockchain improving internet of things?" August 2022,
https://www.plainconcepts.com/Blockchain-iot/

[20] "The benefits of combining the blockchain with IoT,"
https://www.researchgate.net/figure/The-benefits-of-combining-

the-blockchain-with-IoT-8_fig4_350498891

[21] https://www.researchgate.net/figure/Applications-of-Blockchain-for-IoT_fig1_339061504

[22] A. Banafa, "IoT and blockchain convergence: Benefits and challenges," January 2017, https://iot.ieee.org/articles-publications/newsletter/january-2017/iot-and-blockchain-convergence-benefits-and-challenges.html

[23] M. H. Miraz, Blockchain of Things (Bcot): The Fusion of Blockchain and Iot Technologies. Springer Singapore, 2020.

[24] B. Bhushan et al. (eds.), Blockchain Technology Solutions for the Security of IoT-Based Healthcare Systems. Elsevier, 2023.

[25] K. D. Gupta et al. (eds.), Recent Advances in IoT and Blockchain Technology. Bentham Science Publishers, 2022.

[26] G. S. Thakur and H. Patel (eds.), Blockchain Applications in IoT Security. IGI Global, 2020.

[27] B. Krishnan et al. (eds.), Blockchain for IoT. Boca Raton, FL: CRC Press, 2022.

[28] A. A. Elngar et al. (eds.), Convergence of Internet of Things and Blockchain Technologies. Springer International Publishing, 2021.

[29] K. Saini (ed.), Blockchain and IoT Integration: Approaches and Applications. Boca Raton, FL: CRC Press, 2021.

[30] G. C. Deka, P. Zhang, and S. Kim, Role of Blockchain Technology in IoT Applications. Elsevier Science, 2019.

[31] A. Banafa, Secure and Smart Internet of Things (IoT): Using Blockchain and AI (River Publishers Series in Information Science and Technology). ┐ River Publishers, 2018.

[31] D. De, S. Bhattacharyya, and J. J. P. C. Rodriques (eds.), Blockchain based Internet of Things (Lecture Notes on Data Engineering and Communications Technologies, 112). ┌Springer, 2022.

[32] P. Raj et al. (eds.), Blockchain, Artificial Intelligence, and the Internet of Things: Possibilities and Opportunities (EAI/Springer Innovations in Communication and Computing). Springer, 2021.

[33] M. A. Jan and F. Khan (eds.), Application of Big Data, Blockchain, and Internet of Things for Education Informatization: First EAI International Conference, BigIoT-EDU 2021, Virtual Event, August 1–3. Springer, 2021.

[34] V. Upadrista, IoT Standards with Blockchain: Enterprise Methodology for Internet of Things. Apress, 2021.

[35] E. S. Van Engelen, Emerging Technologies: Blockchain of Intelligent Things to Boost Revenues. Business Expert Press, 2020.

[36] A. Jain et al. (eds.), Sustainable Energy Solutions with Artificial Intelligence, Blockchain Technology, and Internet of Things. Routledge, 2024.

[37] Information R Management Association (ed.), Research Anthology on Convergence of Blockchain, Internet of Things, and Security, VOL 1. IGI Global, 2022.

[38] N. Chilamkurti, T. Poongodi, and B. Balusamy (eds.), Blockchain, Internet of Things, and Artificial Intelligence. Routledge, 2021.

[39] H. Patel, Blockchain Applications in IoT Security. IGI Global, 2021

[40] S. Tamwar (ed.), Blockchain for 5G-Enabled IoT: The new wave for Industrial Automation. Springer, 2021.

[41] G. Ramesh et al. (eds.), Blockchain Technology for IoT and Wireless Communications. Boca Raton, FL: CRC Press, 2024.

[42] M. A. Khan et al. (eds.), Decentralised Internet of Things: A Blockchain Perspective. Springer, 2020.

[43] R. Agrawal, J. M. Chatterjee, and A. Kumar (eds.), Blockchain Technology and the Internet of Things: Challenges and Applications in Bitcoin and Security. Apple Academic Press, 2020.

[44] C. Chakraborty et al. (eds.), Implementation of Smart Healthcare Systems using AI, IoT, and Blockchain (Intelligent Data-Centric Systems). Academic Press Inc., 2022.

[45] I. Singh, M. Mohammadian, and S Lee, Blockchain Technology for IoT Applications. Springer Nature Singapore, 2021.

[46] A. Khanna et al. (eds.), Blockchain Applications in IoT

Ecosystem. Springer International Publishing, 2021.

[47] B. K. Mishra et al. (eds.), The Role of IoT and Blockchain: Techniques and Applications. Apple Academic Press, 2022.

CHAPTER 9

BLOCKCHAIN IN SMART CITIES

"A smart city is an intelligent town that provides enormous possibilities for human growth through art, culture, social, architectural, economic, political, environmental, and scientific flowering with the optimal mix of nature, technology, humanity, and arts."

— Amit Ray

9.1 INTRODUCTION

It is hard to imagine modern life without technology. Technology is gradually becoming an indispensable component of our lives because it does miracles toward the business growth and societal development. One of the latest technological innovations, the Blockchain, is poised to disrupt the various aspects of life. Blockchain is an advanced technology that can play a key role in solving the societal issues of urban management.

Cities around the globe are facing tremendous pressures due to rapid urbanization, which causes many economic, social, and environmental problems, which in turn affect people's quality of life. The number of inhabitants in urban areas is increasing. The concept of "smart city" is a response to solve these urban problems [1]. Today, the construction of smart city has progressed steadily. A city becomes smart by transforming itself to a digital city and runs on data, analytics, Internet of things, cloud computing, artificial intelligence, and machine learning. A typical modern city is shown in Figure 9.1 [2].

Figure 9.1 A typical modern city [2].

Smart cities have been considered the wave of the future. Cities play a major role in economic and social aspects of life worldwide. They collaborate, compete, and evolve together with other cities. As people change cities, cities change them [3].

A smart city is a high-tech urban area that connects people, information and technologies in order to increase life quality. Smart cities are those communities that pursue sustainable economic development through investments in human and social capital and manage natural resources through participatory policies. A smart city monitors the conditions and integrates critical infrastructures such as bridges, tunnels, roads, subways, airports, seaports, and buildings. Components of a smart city include smart people, smart governance, smart homes, smart infrastructure, smart technology, smart economy, smart mobility, smart living, smart parking, smart factory, smart health, smart tourism, and smart environment [4]. Some of these components are illustrated in Figure 9.2 [5].

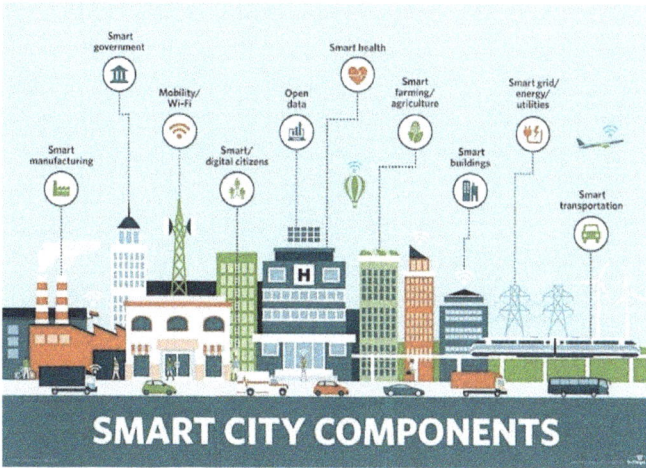

Figure 9.2 The main components of a smart city [5].

A smart city integrates information and communication technology (ICT) in a secure manner so as to manage the city's assets.

Cities around the world are already working to become "smart cities." There are security issues that unfold during the transition from the conventional way of living to the interconnected world of smart cities. The security that Blockchain offers appeals to cities. Blockchain technology can be deployed in any kind of transaction without an intermediator. Cities around the world are using Blockchain technology to enhance their operations and services. It can be used for a variety of transactions between two parties, making it ideal for several municipal applications [6].

This chapter explores the potential of Blockchain (BC) as a technology for enabling smart cities. It begins with providing an overview on Blockchain so that the chapter is self-contained. It explains the concept of smart cities. It discusses some applications of Blockchain in smart cities. It highlights the benefits and challenges of BC in smart cities. The last section concludes with comments.

9.2 OVERVIEW ON BLOCKCHAIN

The term "Blockchain" refers to the way BC stores transaction data – in "blocks" that are linked together to form a "chain." The chain grows as the number of transactions increases. Since every entry is stored as a block on a chain, the care you receive is added to your personal ledger.

At its core, Blockchain is a distributed system recording and storing transaction records. In a Blockchain system, there is no central authority. Instead, transaction records are stored and distributed across all network participants. Rather than having a centrally located database that manages records, the database is distributed to the networks and transactions are kept secure via cryptography. BC eliminates the need for a middleman that traditionally may facilitate such transactions. Unlike traditional trading systems, no intermediary is needed to track the exchange; all parties deal directly with each other [7]. Figure 9.3 shows a model of Blockchain architecture [8], while Figure 9.4 shows how Blockchain works [9].

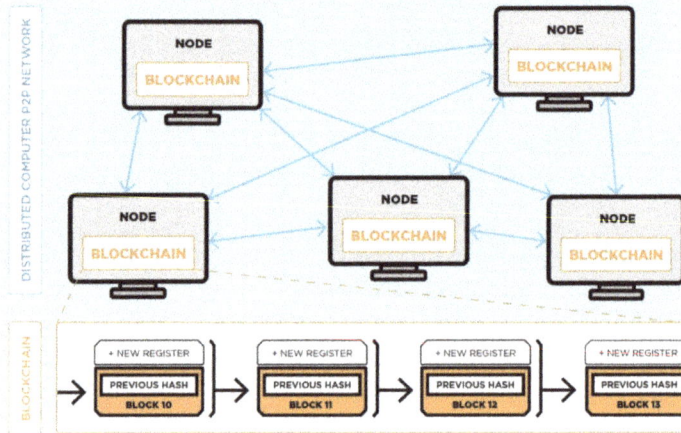

Figure 9.3   A model of Blockchain architecture [8].

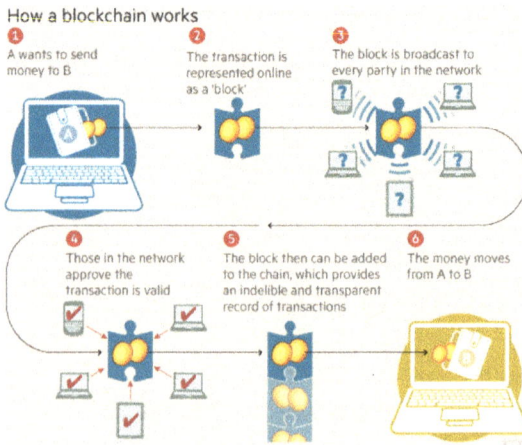

Figure 9.4   How Blockchain works [9].

The Blockchain was designed so transactions are immutable, i.e. they cannot be deleted. Thus, Blockchains are secure and meddle-free by design. Data can be distributed, but not copied. When it comes to digital assets and transactions, you can put almost anything on a Blockchain. Different scenarios call for different Blockchains.

The BC technology currently has the following features [10-12]:

1. *Peer-to-Peer (P2P) Network*: The first requirement of BC is a network, an infrastructure shared by multiple parties. This can be a LAN at a small scale or the Internet at a large scale. All nodes participating in a BC are connected in a decentralized P2P network. Transactions are broadcast to the P2P network. Due to some limitations of P2P networks, some vendors have provided cloud-based BCs.

2. *Cascaded Encryption*: A BC uses encryption to protect transaction data. Blocks are encrypted in a cascaded manner, i.e. the encryption result of the previous block is used in encrypting the current block. The BC is secured by public key cryptography, with each peer generating its own public-private key pairs.

3. *Distributed Database*: A BC is digitally distributed across a number of computers. Each party on a BC has access to the entire database and no single party controls the data or the information. Since BC is decentralized, there is no need for central authorizes such as banks.

4. *Transparency with Pseudonymity*: Each node or participant on a Blockchain has a unique 30-plus-character alphanumeric address that identifies it. Users can choose to remain anonymous or provide proof of their identity to others.

5. *Irreversibility of Records*: Once a transaction is entered in the database and the accounts are updated, the records cannot be altered. Records on the database is permanent, chronologically ordered, and available to all others on the network.

There are two types of blockhains: public and private. Public Blockchains are cryptocurrencies such as Bitcoin, enabling peer-to-peer transactions. Private Blockchains use Blockchain-based platforms such as Ethereum or Blockchain-as-a-service (BaaS) platforms running on private cloud infrastructure. A private BC is an intranet, while a public BC is the Internet. Companies will be disrupted the most by public Blockchains.

BCs may be permissioned or permissionless. In a permissioned BC, each participant has a unique identity. Permissionless BCs are appealing because they allow anyone to join, participate or leave the protocol execution without seeking permission from a centralized or distributed authority. However, permissionless BCs, such as Ethereum or Bitcoin, face transaction volume constraints. Both permisisoned and permisionless can be implemented in healthcare [13].

9.3 CONCEPT OF SMART CITIES

The majority of world's population resides in cities. Over 90 percent of urban growth is occurring in the developing world. Modern cities are monstrous communities, with millions of residents. They are the economic engines of the modern world because they generate economic opportunities. Cities bring individuals together and foster interchange of information by people of different cultures and skills.

A smart city incorporates information and communication technology (ICT) to enhance the quality of life of its citizens and improve the efficiency of services such as energy, water, and transportation. It covers a wide range of scenarios for citizens in their everyday life, such as smart home, smart traffic, smart factory, smart environment and smart agriculture, smart tourism, and smart healthcare. To do all this requires digital infrastructures that can interconnect the sensors, devices, and machines that constitute the public systems. However, the more connected devices are, the more vulnerable they are cyberattacks. Blockchain can ease some of those worries.

Right now, there are very few cities that can claim to be truly "smart" to the core, e.g. Songdo in South Korea and Dubai (which aims to be a global leader in the smart economy or the first Blockchain-powered government in the world by 2020). The major challenge hindering the rise of the smart city is not technology but the data generated by this technology and how we keep it secure. This will be the single most important priority for the city government. This is where Blockchain (BC) comes in since it has the potential to provide this security. Blockchain would enable local governments to streamline city data and make it un-hackable [10]. The Blockchain, when combined with big data technology, can allow multiple parties to interact cooperatively. It can help city government achieve their digital economy goals while ensuring maximum network security. In smart cities, the government has the responsibility of governance, social issues, mobility, security, etc.

Several cities worldwide are embarking on Blockchain related initiatives, such as those in Australia, China, Denmark, United Arab Emirates, Estonia, Georgia, Ghana, Honduras, Malta, Russia, Sweden, Singapore, Spain, Switzerland, United Kingdom, Ukraine, and United States [14]. Perhaps the best-known example of smart city integration is Dubai, where the government is leveraging several emerging technologies including blockchain, IoT and AI to make Dubai "the happiest city on earth" and the first city fully powered by blockchain by 2021. Today, Dubai is regarded as one of the most digitally progressive cities in the world [15]. Dubai is the capital of the United Arab Emirates and the most populous city in the country.

9.4 BLOCKCHAIN IN SMART CITIES

Blockchain technology supports various applications in many sectors including industries, healthcare, automotive, finance, manufacturing, education, and governments. It is boosting the bottom line of private and public sectors. A smart city based upon blockchain is shown in Figure 9.5 [8]. An application of Blockchain technology to smart cities is illustrated in Figure 9.6 [1].

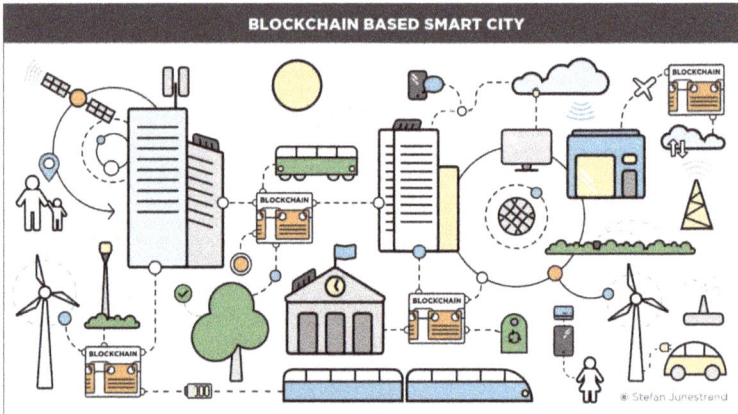

Figure 9.5 A smart city based upon blockchain [8].

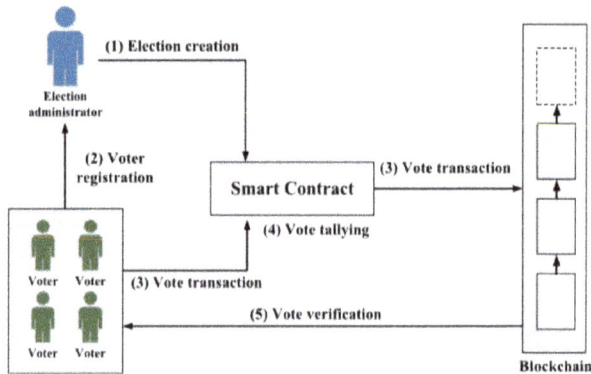

Figure 9.6 Applying Blockchain technology to smart cities [1].

The potential applications of Blockchain technology within the smart city include the following [16]:

- *City Government*: Upgrading to Blockchain would make it easier for city governments to access and manage data. Blockchain technology can help governments boost security in the financial system. Citizen participation and voting system can be enhanced by Blockchain. The application of Blockchain in e-governance is gaining ground towards the concept of e-democracy. The police department can use the Blockchain for public data safety. Some governments have pioneered   Blockchain application in their cities.

Smart contracts can be used for digitizing citizen rights and identification, transparent voting, tax, track ownership of assets, remove paper, and automate bureaucratic processes. A number of cities, governments, and corporations around the world have embraced blockchain, piloting blockchain initiatives. Figure 9.7 shows countries and cities with Blockchain-based government [17].

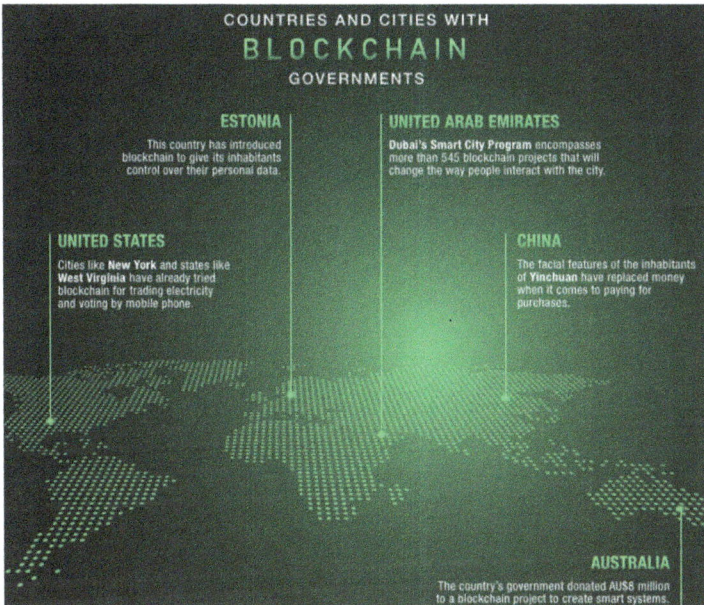

Figure 9.7   Countries and cities with Blockchain-based government [17].

For example, Dubai with IBM is moving toward introducing Blockchain technology in city services. Blockchain e-voting will be "publicly viewable and transparent."

• *City Residents*: This concept of data portability plays a major role in how the city improves the lives of its residents. With permission-based Blockchain, residents have full control over their data and can decide who to share the data with. The residents have full faith in the government through cooperation between the government and the residents. Blockchain can promote the inclusion of persons with disability.

• *Transportation*: Most cities rely heavily on public transportation. In addition to software and diagnostics that improve vehicle safety and performance, technological advances now allow vehicles to detect traffic patterns. In the transport industry, Blockchain technology can help mitigate the risks of cybersecurity by providing a trusted and secure platform.

- *Education*: This is the collective pursuit of truth and the transfer of knowledge across generations. It is a global community characterized by consensus, transparency, and permanence. Like education, Blockchain is intended to transfer not just content, but also the value inherent in that content. Blockchain is the next technological chapter in a long trend of decentralization in the higher education sector. Blockchain offers a model for the Blockchain's ability to manage, share, and protect digital content makes it ideal for helping researchers, faculty members, and other higher-ed principals create intellectual property and share it [18].

- *Healthcare*: This is a prominent area which could benefit greatly from Blockchain. Health is fundamental to the happiness of citizens. The decentralized ledger can be used to store personal details of the patient. The main motivation for using Blockchain in healthcare is to solve the data integrality, data interoperability, and privacy issues in current health systems. Blockchain eliminates the need of a middleman who plays the role of verifying transaction in the healthcare industry [19].

- *Smart Services*: Blockchain can make a smart city even smarter. It can provide smart services such as smart education, smart administration, smart electronic voting, smart mobility and transport, smart home, smart contracts, smart infrastructure, smart energy, smart factory, smart payment, smart currency, etc. and improve of public services in the city. Some of these smart services are illustrated in Figure 9.8 [20].

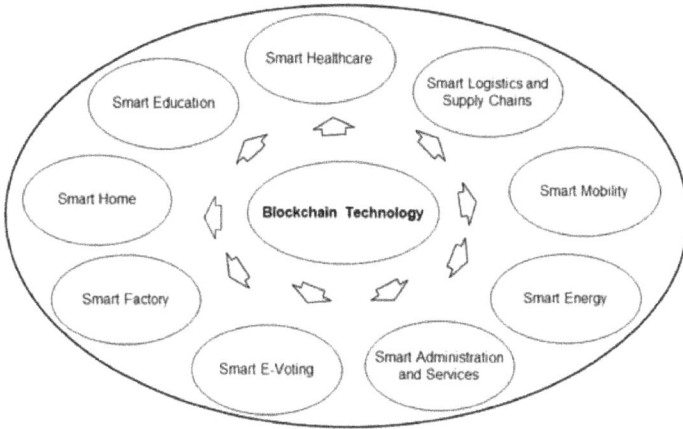

Figure 9.8 Blockchain provides smart services [20].

Blockchain technology is also transcending in the domain of education and improving the extent of educational facilities to remote areas. Blockchain can also be a precursor to smart transport systems which can help in improving mobility across the urban centers of the world. Blockchain technology can expedite the process of smart homes.

• *Media Sector*: The use of blockchain can help media industry in securing creative and intellectual property rights. Due to secure, disintegrated network it will help media and creative groups to authenticate and attribute the original source. Digital transactions, smart contracts, electronic signatures, micro metering are few things which will help the segment tremendously.

Other applications include real estate, water management, insurance industry, notary services, digital identities, supply chain, trade logistics, finance management, registry businesses, money transfers, environment, tourism, and energy.

9.5 BENEFITS

The Blockchain is unquestionably a powerful technology set to change the way transactions are managed and executed. It can provide a great impetus in the development of smart cities. Its major benefit is that it can never get vulnerable to hacking. It is the new business operating system. It is a transparent, neutral, non-hierarchical, accessible, and secure. There is no single point of failure. The use of Blockchain technology in smart cities is quite promising. It is a good instrument for ruling out corruption and inefficiency from the smart cities. Other benefits include [15,21,22]:

- *Improved Cybersecurity*: Cybercrimes are increasing daily. Blockchain can prove helpful in reducing the risk of cyberattacks. End-to-end encryption, secure communication and authentication made possible due to Blockchain can help improve cybersecurity with AI and IoT devices. The integrity of the updates can be verified using Blockchain to prevent malicious software from being installed.

- *Enhanced Healthcare*: Blockchain sees a wide range of applications in the healthcare sector. It is being used to create a distributed system for patient health records and transparent supply chains for medicines as well as manage the outbreak of harmful diseases. Blockchain for smart cities can be combined with other advanced technologies like AI to further improve the healthcare sector.

- *Better Waste Management*: As global urbanization continues, resources will become strained. A smarter city uses technology to transform its core systems and optimize limited resources. Blockchain improves the efficiency of the entire waste management process by using IoT sensors and AI prediction modeling. It can provide real-time tracking of various aspects related to waste management. It can provide transparent, immutable information regarding the amount of waste collected, who collected it, and how the waste is being recycled or disposed of.

- *Simplified Education*: Blockchain for smart cities can simplify the education processes. Educational institutions have to deal with a tremendous amount of student data. Similarly, transferring the data between multiple institutions becomes tedious and time-consuming.

- *Increased Energy Savings*: Resource conservation is a primary focus while developing future smart cities with

Blockchain technology. Blockchain can be used to increase energy savings in smart cities.

• *Efficient Mobility*: Blockchain can help give a boost to transportation services in smart cities. Using Blockchain for smart cities can help create secure vehicle owner data. Government agencies can know which citizens use their cars daily and offer them discounts to encourage them to take public transport.

• *Enhance Security*: The data records in Blockchain are encrypted end-to-end and cant be altered. It helps to prevent fraud and unauthorized activity. Blockchain is in the best position to provide the needed security of private information and the ability to share data only with authorized parties. Rather than being stored in a single server, information is stored across a network of computers, making it difficult for hackers to view data.

• *Stability*: With blockchain there is no single point of failure and therefore the entire system is more resistant to cyberattacks or just technical problems caused by natural hazards.

• *Interoperability*: Cities' ecosystem players (municipalities, utilities, retailers, etc.) should communicate with each other in making decisions. With so many systems, interoperability can quickly become unwieldy. Blockchain allows the use of fewer digital ledgers, therefore reducing intermediaries and costly interoperability problems.

• *Trust*: A lot of important private information is stored and handled across the city's ecosystem. Data that directly relates to the health and wellbeing of citizens is made publicly available.

• *Better Transparency*: Blockchain provides full transparency by maintaining a complete history of past transactions within the network. It gives access to the user to track the data. Blockchain is a peer-to-peer network that makes it highly transparent, and anyone can trace back the transactions made.

• *Improved Traceability*: Blockchain improves traceability

by recording all the transactions without a single network owner. It is possible to share provenance data directly with customers, which helps combat issues like counterfeit goods, compliance violations, delays, and waste.

• *Increased Efficiency*: Blockchain enables the completion of transactions faster and more efficiently. It also enhances energy efficiency

• Automation: Along with Blockchain, smart contracts eliminate third parties, increase efficiency, and further accelerate the automation process.

• *Energy Savings*: Blockchain can increase energy savings in smart cities by establishing a peer-to-peer energy trading platform where citizens can use their surplus electricity and pay subscribers who need additional energy.

• *Scalability*: As our world becomes increasingly digital, we need technology that can scale it while providing the necessary security, trust, and accountability. It is believed that blockchain might be is that technology.

9.6 CHALLENGES

The benefits mentioned earlier are contingent to some challenges in practice. Privacy and security are major issues in Blockchain because of the distributed nature of node connections. Building Blockchains for every single transaction in a large city is not currently feasible. The complete realization of a Blockchain that can truly harness the full potential of smart cities is hampered by the speed and scalability of the Blockchain technology. Blockchain will not be standalone technology in smart cities; instead it will be combined with IoT, AI, cloud computing, and other technologies to produce new urban ecosystems founded on collaboration, sharing, sustainability and innovation. Other challenges include the following [23]:

• *New Technology*: The usage of BC in smart cities is still in the developing stage and face many unresolved problems. To tackle problems regarding verification processes, transaction costs, data limits, it needs constant developments.

• *Security and Privacy Issues*: Even though Blockchain

technology ensures secure or private transactions, there are still many loopholes which needs to be addressed regarding strong encryption, transaction details, personal data sharing, etc.

- *Regulatory System*: Monetary system of any country is a domain of its national government. An effective regulatory body along with the help of financial institutions can help governments in tackling bitcoins misuse. Like any new and innovative technology, blockchain needs support from respective governments.
- *Public Awareness*: The technology is yet to reach common masses, especially in developing nations. People are accustomed to age old currency systems. It would take lot of time to shift their focus towards a more decentralized, online based currency system.
- *Initial Costs*: Though blockchain would minimize cost and time in transaction process, but to make an initial shift require lot of capital and resources. Also blockchain consumes lot of power as miners constantly work to validate transactions.
- *Outdated Systems*: Cities will be faced with new challenges, and outdated systems will be taxed in sectors as diverse as housing, utilities, employment, property, technology, healthcare, and many more.

9.7 CONCLUSION

Blockchain is a peer-to-peer distributed ledger technology which records transactions, agreements, contracts, and sales. It is an emerging technology has several useful features. It is regarded as one of the most disruptive technologies of our time. Although Blockchain is still in its infancy, it may soon become as ubiquitous as electricity. The governments, tech companies, and financial institutions could all see how Blockchain could revolutionize the way we conduct transactions in the smart cities of the future, establishing trust and transparency in government. Applying Blockchain technology to smart cities produces several benefits such as trust-free, transparency, pseudonymity, democracy, automation, decentralization, and security. Today, it is hard to imagine smart cities without Blockchain.

To bring together quality of life, economic growth and environmental sustainability, cities of any size are embracing digital technologies. The first smart cities are already being created, and cities need to be prepared to become 100% smart cities. Soon, every one of us can be called smart citizens. There are many solutions that utilize blockchain technology for cities of the future. Blockchain is emerging as one of the most appealing smart city technologies that have the potential to make smart city operations more transparent, secure, efficient, and resilient. It has the potential to become an underlying operating system that governs the way our cities function in the future. More information about Blockchain in smart cities can be obtained from the books in [24-34] and the following related journals:

- *Cities*
- *Smart Cities*

REFERENCES

[1] J. Xie et al., "A survey of Blockchain technology applied to smart cities: Research issues and challenges," IEEE Communications Surveys & Tutorials, vol. 21, no. 3, Third Quarter 2019, pp. 2795-2830.

[2] "How blockchain can empower smart cities - and why interoperability will be crucial," April 2021,

https://www.weforum.org/agenda/2021/04/how-blockchain-can-empower-smart-cities-gtgs21/

[3] M. N. O. Sadiku, A. E. Shadare, E. Dada, and S. M. Musa, "Smart cities," International Journal of Scientific Engineering and Applied Science, vol. 2, no. 10, Oct. 2016, pp. 41-44.

[4] V. Albino, U. Berardi, and R. M. Dangelico, "Smart cities: Definitions, performance, and initiatives," Journal of Urban Technology, vol. 22, no. 1, 2015, pp. 3-21

[5] N. Mittal " Blockchain's role in developing smart cities," March 2019,

https://theBlockchainland.com/2019/03/05/Blockchain-role-developing-smart-cities/

[6] M. N. O. Sadiku, Y. P. Akhare, A. Ajayi-Majebi, and S. M. Musa, "Blockchain in Smart Cities," International Journal of Trend in Research and Development, vol. 7, no. 4, August 2020.

[7] M. N. O. Sadiku, Y. Wang, S. Cui, and S. M. Musa, "A Primer on Blockchain," International Journal of Advances in Scientific Research and Engineering, vol. 4, no. 2, February 2018, pp. 40-44.

[8] S. Junestrand, "A blockchain-based governance model for public services in smart cities," October 5, 2018

https://www.openaccessgovernment.org/a-blockchain-based-governance-model/52928/

[9] E. Tarasenko, "Blockchain in government," February, 2023,

https://merehead.com/blog/blockchain-in-goverment/

[10] M. Winterson, "Opinion: Is Blockchain the breakthrough smart cities need?" September 2018,

https://www.arabianbusiness.com/technology/405261-abe-1935-is-

Blockchain-the-breakthrough-smart-cities-need

[11] M. Iansiti and K. R. Lakhani, "The truth about Blockchain," Harvard Business Review, Jan./Feb. 2017.

https://hbr.org/2017/01/the-truth-about-Blockchain

[12] W. T. Tsai et al., "A system view of financial Blockchains," Proceedings of IEEE Symposium on Service-Oriented System Engineering, 2016, pp. 450-457.

[13] Z. Alhadhrami et al., "Introducing Blockchains for healthcare," Proceedings of

International Conference on Electrical and Computing Technologies and Applications, 2017.

[14] C. Shen and F. Pena-mora, "Blockchain for cities—A Systematic literature review," IEEE Access, vol. 6, November 2018, pp. 76787-76819.

[15] "Blockchain in smart cities: a reality check," February 2020,

https://blog-idceurope.com/blockchain-in-smart-cities-a-reality-check/

[16] "Blockchain integral to regional smart cities, digital economy, says Booz Allen report," July 2016,

https://www.khaleejtimes.com/business/local/Blockchain-integral-to-regional-smart-cities-digital-economy-booz-allen

[17] "Blockchain technology at the service of urban management,"

https://www.iberdrola.com/innovation/blockchain-for-smart-cities-urban-management#:~:text=Blockchain%20allows%20both%20the%20public,used%20without%20compromising%20people's%20privacy.

[18] " 5 Ways Blockchain is revolutionizing higher education," January 2019,.

https://www.forbes.com/sites/oracle/2019/01/02/5-ways-Blockchain-is-revolutionizing-higher-education/#2507ef4c7c41

[19] J. Qiu, "Towards secure and smart healthcare in smart cities using Blockchain,"

Proceedings of the IEEE International Smart Cities Conference, September 2018,.

[20] H. Treiblmaier, A. Rejeb, and A. Strebinger, "Blockchain as a driver for smart city development: Application fields and a comprehensive research agenda," Smart Cities, vol. 3, no. 3, August 2020, pp. 853-872.

[21] N. Joshi, "6 Ways in which Blockchain makes your smart city even smarter,"

https://www.forbes.com/sites/naveenjoshi/2022/04/07/6-ways-in-which-Blockchain-makes-your-smart-city-even-smarter/?sh=1372037b7f5d

[22] "Blockchain uses cases in smart cities," January 2023,

https://itsavirus.com/updates/blockchain-uses-cases-in-smart-cities

[23] "Blockchain in smart cities," August 2017, Unknown Source.

[24] V. Kumar et al. (eds.), Smart City Infrastructure: The Blockchain Perspective. Wiley-Scrivener, 2022.

[25] M. Pustišek, N. Živić, and A. Kos, Blockchain: Technology and Applications for Industry 4.0, Smart Energy, and Smart Cities. De Gruyter, 2021.

[26] S. Krishnan et al. (eds.), Blockchain for Smart Cities. Elsevier, 2021.

[26] D. Singh and N. S. Rajout (eds.), Technology for Smart Cities. Springer, 2021.

[27] R. Kumar et al. (eds.), Convergence of IoT, Blockchain, and Computational Intelligence in Smart Cities. Boca Raton,FL: CRC Press, 2024.

[28] J. Gao et al., Smart Cities: Blockchain-Based Systems, Networks, and Data. CRC Press, 2022.

[29] S. Krishnan, R. Kumar, and V. Balas (eds.), Green Blockchain Technology for Sustainable Smart Cities. Elsevier, 2023,

[30] S. Mahankali, Blockchain & The Smart Cities: Lessons From Singapore, the World's Smartest Digital Nation. Notion Press, 2020.

[31] P. Swarnalatha and S. Prabu (eds.), Blockchain Technologies for Sustainable Development in Smart Cities. Engineering Science Reference, 2022.

[32] C. G. Reddick, M. P. Rodríguez-Bolívar, and H. J. Scholl (eds.), Blockchain and the Public Sector: Theories, Reforms, and Case Studies. Springer, 2021.

[33] E. Estevez, T. A. Pardo, and H. J. Scholl (eds,), Smart Cities and Smart Governance: Towards 22nd Century Sustainable City. Springer, 2021.

[34] H. Toraman, The Blockchain in Smart Cities. Analysis of Blockchain Use Cases for Smart Mobility. GTIN Verlag, 2021.

CHAPTER 10

BLOCKCHAIN IN CYBERSECURITY

"If you think technology can solve your security problems, then you don't understand the problems and you don't understand the technology."

— Bruce Schneier

10.1 INTRODUCTION

Today, data is a new oil for any business growth. As enterprises amass tons of sensitive data, they require means of storing and processing the data smartly and securely. In other words, as the world goes online it is imperative to maintain the security of online data. Any attempt to gain unauthorized access to computers or online data with the intent to cause harm is a cyber attack. The growing reliance on digital services has led to an escalation in cyber risks and attacks specifically targeting banks and financial institutions.

Cybercriminals are increasing the frequency and sophistication of cyber attacks by pooling their knowledge and leveraging new technologies. Cybercrime costs the global economy an estimated $450 billion every year. Our current security protocols simply cannot keep up with the relentless and clever attacks.

Cybersecurity is the practice of protecting systems and networks from digital attacks. Blockchain is the latest cybersecurity technology that is gaining popularity and recognition. Blockchain provides security, anonymity, and data integrity without the need of a third party. Blockchain is regarded as a new weapon in cybersecurity. Cybersecurity is built into Blockchain technology because of its inherent nature of being a decentralized system based on principles of security, privacy, and trust. In today's digital world, where cyber threats are a major concern, the use of Blockchain technology can provide a robust solution to enhance cybersecurity [1].

Cybersecurity is a major concern for business in the digital environment. With the increase in cyber attacks, organizations seek innovative methods to secure their data and assets. Blockchain technology is emerging as the ultimate weapon in the fight against cybercrime. It is a revolutionary technology poised to change the future of computing and disrupt several industries with innovative solutions. By leveraging Blockchain-based storage solutions that provide decentralized storage capability, organizations can protect their digital information and assets. The decentralized nature of Blockchain means that it is not controlled by a single entity, making it more resistant to attacks. The popularity of Blockchain has increased worldwide due to its lasting impact on the world [2]. Blockchain is associated with benefits including high level of transparency, integrity, trust, and confidence for the participants.

This chapter provides a primer on Blockchain security. It begins with providing an overview of on Blockchain. It covers cybersecurity basics. It discusses Blockchain security and some of its applications. It highlights the benefits and challenges of Blockchain in cybersecurity. The last section concludes with comments.

10.2 OVERVIEW OF BLOCKCHAIN

Blockchain (BC) technology is a permanent record of online transactions. It is a distributed tamper-proof database, shared, and maintained by multiple parties. It is a new enabling technology that is expected to revolutionize many industries, including business. It has the potential for addressing significant business issues. The BC technology allows participants to move data in real-time, without exposing the channels to theft, forgery, and malice.

The term "Blockchain" refers to the way BC stores transaction data – in "blocks" that are linked together to form a "chain." The chain grows as the number of transactions increases. Since every entry is stored as a block on a chain, the care you receive is added to your personal ledger. The first Blockchain was conceived in 2008 by an anonymous person or group known as Satoshi Nakamoto, who published a white paper introducing the concept of a peer-to-peer electronic cash system he called Bitcoin [3]. A typical Blockchain architecture is shown in Figure 10.1 [4]

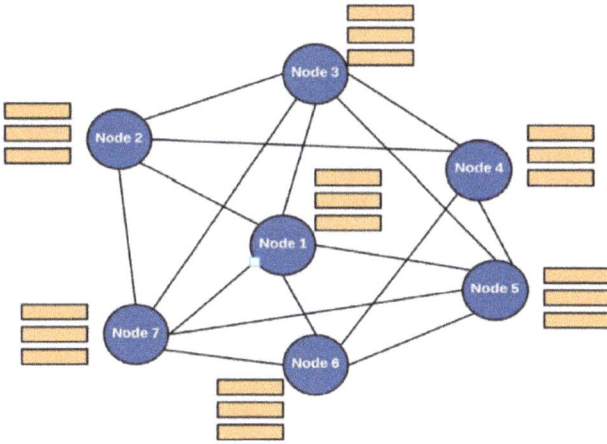

Figure 10.1   The Blockchain architecture [4].

At its core, Blockchain is a distributed system recording and storing transaction records. In a Blockchain system, there is no central authority.   Instead, transaction records are stored and distributed across all network participants. Rather than having a centrally located database that manages records, the database is distributed to the networks and transactions are kept secure via cryptography. BC eliminates the need for a middleman that traditionally may facilitate such transactions. Figure 10.2 shows how Blockchain works [5].

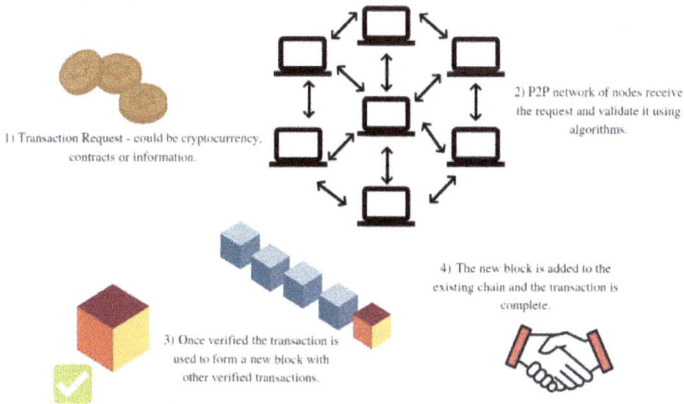

Figure 10.2   How the Blockchain works [5].

Fundamentally, Blockchains are distributed digital database that record and maintain a list of transactions taking place in real time. They may also be regarded as decentralized ledgers that sequentially record transactions or interactions among users within a distributed network. They have the following properties [6]:

- Firstly, they are autonomous. They run on their own, without any person or company in charge.

- Secondly, they are permanent. They are like global computers with 100 percent uptime. Because the contents of the database are copied across thousands of computers, if 99 per cent of the computers running it were taken offline, the records would remain accessible and the network could rebuild itself.

- Thirdly, they are secure and tamper-proof. Each record in Blockchain is time stamped and stored cryptographically. The encryption used on Blockchains like Bitcoin and Ethereum is industry standard, open source, and has never been broken.

- Fourthly, they are open, allowing anyone to develop products and services on them.

- Fifthly, as Blockchain is a shared system, costs are also shared between all of its users.

The Blockchain was designed so transactions are immutable, i.e. they cannot be deleted. Thus, Blockchains are secure and meddle-free by design. Data can be distributed, but not copied. When it comes to digital assets and transactions, you can put almost anything on a Blockchain. Different scenarios call for different Blockchains. Blockchain is used in different areas such as depicted in Figure 10.3 [7].

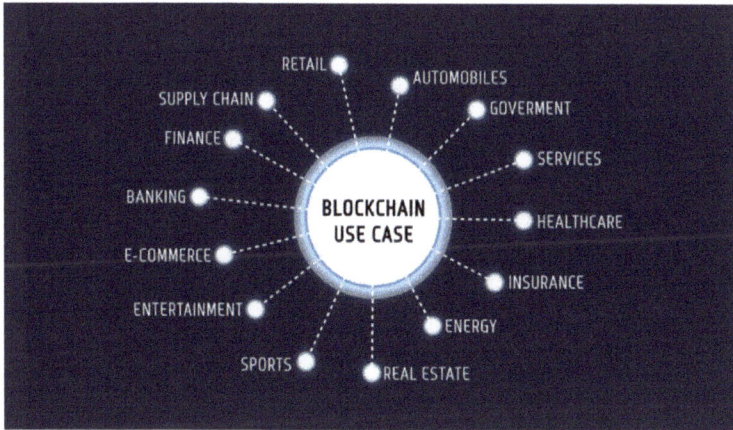

Figure 10.3 Different uses of Blockchain [7].

The BC technology currently has the following features [8,9]:

1. *Peer-to-Peer (P2P) Network*: The first requirement of BC is a network, an infrastructure shared by multiple parties. This can be a LAN at a small scale or the Internet at a large scale. All nodes participating in a BC are connected in a decentralized P2P network. Transactions are broadcast to the P2P network. Due to some limitations of P2P networks, some vendors have provided cloud-based BCs.

2. *Cascaded Encryption*: A BC uses encryption to protect transaction data. Blocks are encrypted in a cascaded manner, i.e. the encryption result of the previous block is used in encrypting the current block. The BC is secured by public key cryptography, with each peer generating its own public-private key pairs.

3. *Distributed Database*: A BC is digitally distributed across a number of computers. Each party on a BC has access to the entire database and no single party controls the data or the information. Since BC is decentralized, there is no need for central authorizes such as banks.

4. *Transparency with Pseudonymity*: Each node or participant on a Blockchain has a unique 30-plus-character alphanumeric address that identifies it. Users can choose to remain anonymous or provide proof of their identity to others.

5. *Irreversibility of Records*: Once a transaction is entered

in the database and the accounts are updated, the records cannot be altered. Records on the database is permanent, chronologically ordered, and available to all others on the network.

There are two types of Blockchains: public and private. Public Blockchains are cryptocurrencies such as Bitcoin, enabling peer-to-peer transactions. Private Blockchains use Blockchain-based platforms such as Ethereum or Blockchain-as-a-service (BaaS) platforms running on private cloud infrastructure. A private BC is an intranet, while a public BC is the Internet. Companies will be disrupted the most by public Blockchains.

10.3 CYBERSURITY BASICS

Cybersecurity refers to a set of technologies and practices designed to protect networks and information from damage or unauthorized access. It is vital because governments, companies, and military organizations collect, process, and store a lot of data. As shown in Figure 10.4, cybersecurity involves multiple issues related to people, process, and technology [10].

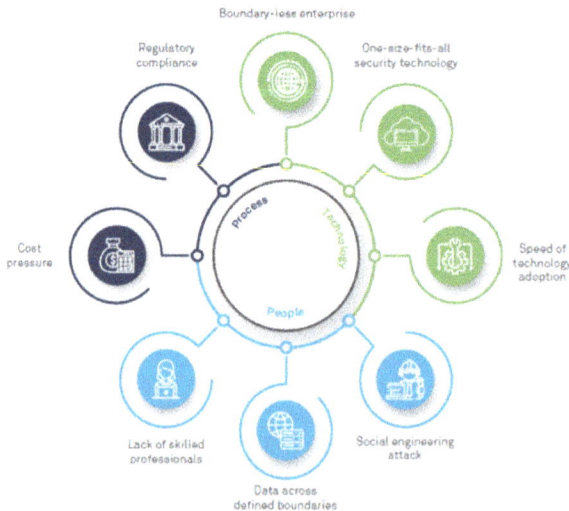

Figure 10.4 Cybersecurity involves multiple issues related to people, process, and technology [10].

A typical cyber attack is an attempt by adversaries or cybercriminals to gain access to and modify their target's computer system or network. Cyber attacks are becoming more frequent, sophisticated, dangerous, and destructive. They are threatening the operation of businesses, banks, companies, and government networks. They vary from illegal crime of individual citizen (hacking) to actions of groups (terrorists) [11].

The cybersecurity is a dynamic, interdisciplinary field involving information systems, computer science, and criminology. The security objectives have been availability, authentication, confidentiality, nonrepudiation, and integrity. A security incident is an act that threatens the confidentiality, integrity, or availability of information assets and systems [12].

- *Availability*: This refers to availability of information and ensuring that authorized parties can access the information when needed. Attacks targeting availability of service generally leads to denial of service.

- *Authenticity*: This ensures that the identity of an individual user or system is the identity claimed. This usually involves using username and password to validate the identity of the user. It may also take the form of what you have such as a driver's license, an RSA token, or a smart card.

- *Integrity*: Data integrity means information is authentic and complete. This assures that data, devices, and processes are free from tampering. Data should be free from injection, deletion, or corruption. When integrity is targeted, nonrepudiation is also affected.

- *Confidentiality*: Confidentiality ensures that measures are taken to prevent sensitive information from reaching the wrong persons. Data secrecy is important especially for privacy-sensitive data such as user personal information and meter readings.

- *Nonrepudiation*: This is an assurance of the responsibility to an action. The source should not be able to deny having sent a message, while the destination should not deny having received it. This security objective is essential for accountability and liability.

Everybody is at risk for a cyber attack. Cyber attacks vary from illegal crime of individual citizen (hacking) to actions of groups (terrorists). The following are typical examples of cyber attacks or threats [13]:

- *Malware*: This is a malicious software or code that includes traditional computer viruses, computer worms, and Trojan horse programs. Malware can infiltrate your network through the Internet, downloads, attachments, email, social media, and other platforms. Spyware is a type of malware that collects information without the victim's knowledge.

- *Phishing*: Criminals trick victims into handing over their personal information such as online passwords, social security number, and credit card numbers.

- *Denial-of-Service Attacks*: These are designed to make a network resource unavailable to its intended users. These can prevent the user from accessing email, websites, online accounts or other services.

- *Social Engineering Attacks*: A cyber criminal attempts to trick users to disclose sensitive information. A social engineer aims to convince a user through impersonation to disclose secrets such as passwords, card numbers, or social security number.

- *Man-In-the-Middle Attack*: This is a cyber attack where a malicious attacker secretly inserts him/herself into a conversation between two parties who believe they are directly communicating with each other. A common example of man-in-the-middle attacks is eavesdropping. The goal of such an attack is to steal personal information.

These and other cyber attacks are shown in Figure 10.5 [14].

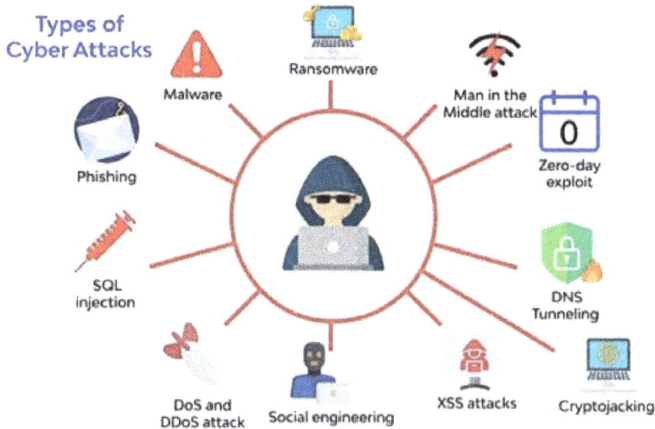

Figure 10.5   Common types of cyber attacks [14].

Cybersecurity involves reducing the risk of cyber attacks. Cyber risks should be managed proactively by the management. Cybersecurity technologies such as firewalls are widely available [15].   Cybersecurity is the joint responsibility of all relevant stakeholders including government, business, infrastructure owners, and users. Cybersecurity experts have shown that passwords are highly vulnerable to cyber threats, compromising personal data, credit card records, and even social security numbers. Governments and international organizations play a key role in cybersecurity issues. Popular applications of cybersecurity are displayed in Figure 10.6 [16].

Figure 10.6   Some applications of cybersecurity [16].

10.4 BLOCKCHAIN SECURITY

Blockchain technology is based on decentralization and encryption. Each user has a private key to add blocks and make changes, and a public key to enable others to access the database so they can observe the modifications. Blockchain offers several opportunities to maintain a high level of data security through reliable data encryption mechanisms, data integrity, network resilience, and scalability. By leveraging the power of Blockchain technology, organizations can ensure that their data are safe from manipulation, unauthorized access, and malicious attacks. Blockchain security deals with assurance services, cybersecurity frameworks, and best practices to mitigate the risk of fraud and cyber-attacks.

As shown in Figure 10.7 [17], Blockchain security relies on three fundamental elements [17,18]:

Figure 10.7   Three fundamental elements of Blockchain security [17].

- *Confidentiality*: Confidentiality refers to the privacy of information stored and processed digitally. It means to ensure that only interested and authorized parties access the appropriate data.

- *Integrity*: Blockchains built-in characteristics of immutability and traceability help organizations ensure data integrity.

- *Availability*: Data remains available through various nodes and thus full copies of the ledger can be accessed at all times. All the Blockchain nodes will have a complete Blockchain database so that if a node is unavailable, it will affect the Blockchains performance.

The CIA triad is an industry standard and all people who operate in the field of cybersecurity must know the three terms.

Blockchain security is a complete risk management system for Blockchain systems. It describes the measures to guard against unwanted access, manipulation, and interruption of a Blockchain network. These measures include network architecture, cryptographic mechanisms, and consensus algorithms. With them, it becomes possible to safeguard the integrity and immutability of a Blockchain network [19].

10.5 APPLICATIONS OF BLOCKCHAIN SECURITY

Different industries can use Blockchains to improve the security of their data, financial transactions, and communication. The industries that can benefit the most from applying the Blockchain for cybersecurity include [20-22]:

- *Finance*: In the finance sector, the biggest value of a Blockchain is in its data immutability and transaction transparency. See Chapter 3 for more.

- *Healthcare*: The healthcare industry endures a constant barrage of cyber attacks. The most common examples of Blockchain implementation in healthcare are related to securely storing and quickly transferring medical data. Blockchain technology could be the badly-needed solution to a problem that puts patients and hospitals at severe risk.

- *Real Estate*: The real estate platforms use Blockchains for solving two major tasks: ensuring safe data storage and automating key processes such as validating property ownership and transferring funds. A Blockchain also offers reliability and automation, which are crucial for the successful operation of real estate businesses.

- *Supply Chain*: A Blockchain can store tamper-proof records of all operations, transactions, and freight data to simplify the analysis of a supply chain's efficiency and operations. Global giants like Walmart, BMW, and FedEx deploy Blockchains for improving data security and operational transparency.

- *Governance*: Blockchains can also be useful for improving the security and transparency of many government processes: tax collection, information governance, elections, etc. During elections, a Blockchain can be used to speed up vote counting and ensure the accuracy of results. For example, the Australian government has plans to develop a cybersecurity network based on Blockchain. China's government is attempting to secure vital government and military information using Blockchain cybersecurity.

- *Digital Identity*: Using Blockchain technology, we have the tools to build identity management systems. Digital IDs may be created both for users and endpoints within an organization. Users can be identified using a combination of first and last name, date of birth, nationality, and social security number. Digital identities may be secured using the principles of private/public-key cryptography. Once the digital identity has been generated, it can then be stored on the chain in an immutable way so that attackers cannot tamper with it.

10.6 BENEFITS

Blockchain technology is not only an innovative technology that revolutionizes the way we store and share data, it is also a powerful cybersecurity tool. It can be used to prevent any data breach, identity theft, cyberattacks, or criminal acts in transactions. The cybersecurity industry can benefit from Blockchain's unique features, which create a virtually impenetrable wall between a hacker and your information. The combination of Blockchain and cybersecurity has intrigued executives and technology experts. Other benefits include [14,16,23]:

- *Safe Data Transfers*: The Blockchain enables fast and secure transactions of data. PKI in Blockchain maintains authentication during data transfers. Smart contracts help to automatically execute agreements between two parties during a transfer.

- *Decentralized Architecture*: One of the main advantages of Blockchain technology is its decentralized architecture. This means that there is no single point of control, making it more difficult for hackers to attack and compromise the system. Blockchain's decentralized and distributed network also helps businesses to avoid a single point of failure.

• *Immutable Records*: One of the popular qualities of Blockchain is its immutability. The data stored on a Blockchain is immutable, meaning that once a transaction is recorded, it cannot be altered or deleted. Blockchain's key feature of immutability and records of any changes to the data help store the data safely and securely. This helps prevent data tampering and ensures the integrity of the data, making Blockchain technology very good at fraud prevention.

• *Tracking and Tracing*: The history of transactions on BC is maintained so that they can be traced anytime. The transaction data is digitally signed by members of the Blockchain network to ensure transparency. All transactions in Blockchains are digitally signed and time-stamped, so that network users can easily trace the history of transactions and track accounts at any historical moment.

• *Confidentiality*: The Blockchain technology provides extensive capabilities for ensuring a user's anonymity. User keys are the only link between a user and their data, and the keys are easy to anonymize. The public key cryptography in a Blockchain network helps maintain the confidentiality of the users. The confidentiality of network members is high due to the public-key cryptography that authenticates users and encrypts their transactions.

• *Fraud Security*: As of today, Blockchains are considered "unhackable," as attackers can impact a network only by getting control of 51% of the network nodes. In the event of a hack, it is easy to define malicious behavior due to the peer-to-peer connections and distributed consensus.

• *Sustainability*: Blockchain systems are decentralized, so the failure of a single node does not affect the entire network This implies that in the case of DDoS attacks, the system will operate as normal thanks to multiple copies of the ledger.

• *Integrity*: BC technology ensures the authenticity and irreversibility of completed transactions. Encrypted blocks contain immutable data that is resistant to hacking. Once a transaction is recorded on the Blockchain, it cannot be altered or deleted. Any changes made to the already recorded data are processed as new transactions.

• *Resilience*: The peer-to-peer nature of the technology ensures that the network will operate round-the-clock even if some nodes are offline or under attack. In the event of an attack, a company can make certain nodes redundant and operate as usual.

• *Smart Contracts*: These programs ensure the execution of contract terms and verify parties. They are self-executing agreements, with terms contained in lines of code, and nearly every Blockchain solution and interaction makes use of them. Blockchain technology can significantly increase the security standards for smart contracts, as it minimizes the risks of cyber-attacks and bugs.

• *Availability*: There is no need to store your sensitive data in one place, as Blockchain technology allows you to have multiple copies of your data that are always available to network users. Having a large number of nodes ensures Blockchain resilience even when some nodes are unavailable.

• *Trust*: Your clients will trust you more if you can ensure a high level of data security. Blockchain technology allows you to provide your clients with information about your products and services instantly.

• *Improved Transparency*: Unless all the participants agree on it, no change can be made on the transaction record. Thus, better transparency and consistency can be assured with Blockchain applications.

• *Efficiency*: We all know that paper-heavy processes are indeed very time-consuming. With Blockchain, you can make this process automated and faster.

• *Cost-efficient*: Using Blockchain technology, not many third parties are required as trusting the information on the Blockchain is enough. Also, you can eliminate reviewing excessive documentation before finalizing a trade. Thus, business operational cost reduction is possible.

10.7 CHALLENGES

In spite of its numerous advantages, Blockchain is not impervious to cybersecurity issues. Blockchain technology is neither perfect nor completely secure. Critics who question the scalability, security, and sustainability of BC technology remain. There will always be people looking for vulnerabilities and means to manipulate the technology in ways the developers never intended. While the Blockchain technology has great potential as a cybersecurity measure, it is also associated with several risks. Other challenges include [14,17,20]:

- *Reliance on Private Keys*: Blockchain relies heavily on private keys for encrypting data. The keys are long sequences of random numbers automatically generated by a wallet. Private keys are used for interacting with the Blockchain and, in contrast to user passwords, cannot be restored. These private keys cannot be recovered once lost.

- *Scalability Challenges*: Scalability may become a constraint when implementing Blockchain, mostly due to block size and response times Blockchain networks have a preset block volume and limits for transactions per second. Integrating Blockchain technology requires a complete replacement of existing systems, which is why companies may find this difficult.

- *Adaptability Challenges*: Although BC technology can be applied to almost any business, some companies may face difficulties integrating it. Blockchain applications can also require complete replacement of existing systems, so companies should consider this before implementing Blockchain technology.

- *Lack of Regulation*: Blockchain concepts are not globally regulated yet. Many countries already have or are working on cryptocurrency regulations. Any Blockchain implementation should be carried out with a close eye on regulatory requirements.

- *Blockchain Literacy*: Learning Blockchain technology requires a profound knowledge of various development, programming languages, and other tools. There are not enough Blockchain developers, Blockchain experts, and cryptography experts.

- *Risk of Cyberattacks*: Blockchain technology greatly reduces the risk of malicious intervention, but it is still not a panacea to all cyber threats. The Blockchain also has its weak spots. If attackers manage to exploit any of these vulnerabilities, it may risk the security of the entire system. A 51% attack occurs when an attacker gains control over more than half of the network's computing power.

- *Theft of Keys*: As secure as a Blockchain may be, things can go wrong if a cybercriminal manages to steal keys.

- *Interoperability*: Weak interoperability limits scalability. From the developer perspective, roadblocks can also be created from platform misconfiguration, communication mistrust, specification errors in application development, and cross-chain smart contract logic problems.

- *High Operating Costs*: The benefits of decentralization come at a technology cost. Blockchain requires high computing power and storage capabilities. This leads to higher costs as compared to non-Blockchain applications.

- *Irreversibility*: There is a risk that encrypted data may be unrecoverable in case a user loses or forgets the private key necessary to decrypt it.

- *Storage Limits*: Each block can contain no more than 1 Mb of data, and a Blockchain can handle only 7 transactions per second in average.

- *Blockchain Literacy*: There are still not enough developers with experience in Blockchain technology and with deep knowledge of cryptography.

- *Lack of Governance*: The operation and use of Blockchain technology in general and distributed ledgers in particular is not well regulated globally. Many countries, including Malta and the US, already have or are working on cryptocurrency regulations.

Some of these challenges are illustrated in Figure 10.8 [20].

Figure 10.8   Some of these challenges of using Blockchain in cybersecurity [20].

In spite of the mixed feelings about Blockchain and its challenges, many industries have invested millions in Blockchain projects

10.8 CONCLUSION

Although Blockchain is a relatively new technology, it seems to be revolutionary. With the maturity of the technology, Blockchain will become far more seamless to adopt as the main guard against cyber threats. Blockchain security is the security measures that aid in keeping the Blockchain network safe and the data stored on it away from unauthorized access, manipulation, and disruption. It is a complete risk management system for Blockchain systems.

Blockchain technology has emerged as a promising solution to address the critical concerns of cybersecurity and data privacy. The adoption of Blockchain technology is taking place at a fast pace. Its potential to enhance cybersecurity is significant. As the technology continues to evolve, it is likely to play an increasingly role in protecting against cyber threats. While it may not be the silver bullet to cybersecurity's problems, Blockchain has great potential to help solve some of the many challenges the industry faces. More information about Blockchain in cybersecurity can be found in the books in [23-37].

REFERENCES

[1] M. N. O. Sadiku, U. C. Chukwu, and J. O. Sadiku, "Blockchain in cybersecurity," Horizon: Journal of Humanities and Artificial Intelligence, vol. 2, no. 9, 2023, pp. 18-28.

[2] P. J. Taylor et al., "A systematic literature review of Blockchain cyber security," Digital Communications and Networks, vol. 6, no. 2, May 2020, pp.147-156.

[3] M. N. O. Sadiku, Y. Wang, S. Cui, and S. M. Musa, "A primer on Blockchain," International Journal of Advances in Scientific Research and Engineering, vol. 4, no. 2, February 2018, pp. 40-44.

[4] M. J. Tuyisenge, "Blockchain technology security concerns: Literature review,"

https://www.diva-portal.org/smash/get/diva2:1571072/FULLTEXT01.pdf

[5] E. Zamani, Y. He, and M. Phillips, "On the security risks of the Blockchain," Journal of Computer Information Systems, vol. 60, no. 6, 2018, pp. 495-506.

[6] S. Depolo, "Why you should care about Blockchains: The non-financial uses of Blockchain technology," March 2016,

https://www.nesta.org.uk/blog/why-you-should-care-about-Blockchains-non-financial-uses-Blockchain-technology

[7] O. Bheda, "What is Blockchain?" https://builtin.com/Blockchain

[8] M. Iansiti and K. R. Lakhani, "The truth about Blockchain," Harvard Business Review, Jan./Feb. 2017.

https://hbr.org/2017/01/the-truth-about-Blockchain

[9] W. T. Tsai et al., "A system view of financial Blockchains," Proceedings of IEEE Symposium on Service-Oriented System Engineering, 2016, pp. 450-457.

[10] "Eliminating the complexity in cybersecurity with artificial intelligence,"

https://www.wipro.com/cybersecurity/eliminating-the-complexity-in-cybersecurity-with-artificial-intelligence/

[11] M. N. O. Sadiku, S. Alam, S. M. Musa, and C. M. Akujuobi,

"A primer on cybersecurity," International Journal of Advances in Scientific Research and Engineering, vol. 3, no. 8, Sept. 2017, pp. 71-74.

[12] M. N. O. Sadiku, M. Tembely, and S. M. Musa, "Smart grid cybersecurity," Journal of Multidisciplinary Engineering Science and Technology, vol. 3, no. 9, September 2016, pp.5574-5576.

[13] "FCC Small Biz Cyber Planning Guide,"

https://transition.fcc.gov/cyber/cyberplanner.pdf

[14] R. Chintawar and M. Sampath, "Blockchain: The weapon for cybersecurity," March 2023,

https://www.encora.com/insights/Blockchain-the-weapon-for-cybersecurity#:~:text=Blockchain%20technology%20can%20be%20used,fraudulent%20activities%20through%20consensus%20mechanisms.

[15] Y. Zhang, "Cybersecurity and reliability of electric power grids in an interdependent cyber-physical environment," Doctoral Dissertation, University of Toledo, 2015.

[16] https://www.researchgate.net/figure/Applications-of-Cybersecurity_fig1_351418658

[17] "Blockchain for cybersecurity: Pros and cons, trending use cases," February 2021,

https://www.apriorit.com/dev-blog/462-Blockchain-cybersecurity-pros-cons

[18] N. L. Lincy and N. Kuriakose, "Cybersecurity aspects of Blockchain,"

International Journal of Creative Research Thoughts, vol. 9, no. 5, May 2021, pp. 310-312.

[19] "Blockchain security v/s cybersecurity: Answering the unanswered!"

January 31, 2023

https://www.immunebytes.com/blog/Blockchain-security-vs-cybersecurity-answering-the-unanswered/

[20] A. Banafa, "Second line of defense for cybersecurity: Blockchain," April 2018,

https://www.bbvaopenmind.com/en/technology/digital-world/
second-line-of-defense-for-cybersecurity-Blockchain/

[21] "Blockchain for cybersecurity: Pros and cons, trending use cases," February 2021,

https://www.apriorit.com/dev-blog/462-Blockchain-cybersecurity-pros-cons

[22] S. Mohan, "Blockchains: Use cases in cybersecurity," February 2022,

https://www.forbes.com/sites/forbestechcouncil/2022/02/02/
Blockchains-use-cases-in-cybersecurity/?sh=4c8ce2335989

[23] M. N. O. Sadiku, Cybersecurity and Its Applications. Moldova, Europe: Lambert Academic Publishing, 2023.

[24] H. E. Poston, Blockchain Security from the Bottom Up: Securing and Preventing Attacks on Cryptocurrencies, Decentralized Applications, NFTs, and Smart Contracts. Wiley, 2022.

[25] K. R. Choo, A. Dehghantanha, and R. M. Parizi (eds.), Blockchain Cybersecurity, Trust and Privacy. Springer, 2020

[26] I. Romdhani, M. Alazab, and Y. Maleh (eds.), Blockchain for Cybersecurity in Cyber-Physical Systems. Springer, 2023.

[27] R. Gupta, Hands-On Cybersecurity with Blockchain:Implement DDoS Protection, PKI-based Identity, 2FA, and DNS Security Using Blockchain. Packt Publishing, 2018.

[28] R. Agrawal and N. Gupta (eds.), Transforming Cybersecurity Solutions using Blockchain. Springer, 2021.

[29] U. P. Rao et al. (eds.), Blockchain for Information Security and Privacy. Auerbach Publications, 2021.

[30] Y. Maleh et al. (eds.), Blockchain for Cybersecurity and Privacy: Architectures, Challenges, and Applications. Boca Raton, FL: CRC Press, 2020.

[31] A. Banafa, 28 Second Line of Defense for Cybersecurity: Blockchain. River Publishers, 2022.

[32] Y. Maleh et al. (eds.), Artificial Intelligence and Blockchain for Future Cybersecurity Applications (Studies in Big Data, 90). Springer, 2021.

[33] Y. Maleh, M. Alazab, and I. Romdhani (eds.), Blockchain for Cybersecurity in Cyber-Physical Systems. Springer, 2023.

[34] S. Mahankali, A. Bhattacharya, and G. B. Alex, Secure Chains
Cybersecurity and Blockchain-powered Automation. BPB PUBN, 2020.

[35] S. Tanwar, Machine Learning, Blockchain, and Cyber Security in Smart Environments: Application and Challenges. Chapman & Hall/CRC Press, 2022.

[36] R. Agrawal, R. Agrawal, and N. Gupta (eds.), Blockchain Applications in Cybersecurity Solutions. Bentham Science Publishers, 2023.

[37] S. Mahankali, A. Bhattacharya, and G. B. Alex, Secure Chains:

Cybersecurity and Blockchain-powered Automation. BPB PUBN, 2020

CHAPTER 11

BLOCKCHAIN IN SOCIAL MEDIA

"Social Media is about the people! Not about your business. Provide for the people and the people will provide for you."

— Matt Goulart

11.1 INTRODUCTION

Modern social networks wield enormous power in society as it is evident by the fact that people are spending significant portions of their lives online. Social media has become a huge part of our life. We turn to it to form groups, foster relationships, and keep in touch with long-distance friends. The social media platform has served as an entry point for establishing connections, content sharing, and social interactions for many users. Social networks play a massive role in our daily communications and interactions. However, centralized control of these platforms has created many problems: data breaches, server outages, de-platforming, censorship, and privacy violations. To combat these issues, developers are building social networks on Blockchain, such as Ethereum [1].

A Blockchain is a decentralized distributed ledger that maintains a record of all transactions on the chain. It is the basis for a new, decentralized version of the internet known as Web 3.0. It makes it possible for users to securely and profitably share content on social media platforms. Web3 proponents think they can build a new system altogether that sidesteps the concentrated power of platforms, a decentralized social media (DeSo) ecosystem where users ideally have more ownership over the content they create.

Social media, such as Facebook, Twitter, Instagram, Youtube, Whatsapp, Snapchat, etc. have changed our way of communication. They have become the center of the modern Internet. Blockchain has emerged as a viable solution for solving many issues, including social media. It is essentially a chain consisting of blocks of information. The main advantage of leveraging Blockchain technology on to social media platforms is its decentralized nature. Blockchain technology offers a promising solution for a new social media architecture. The decentralized nature of Blockchain allows for the creation of platforms where users control their data and have a stake in the network. This eliminates the need for a central authority or intermediary, promoting transparency and fairness [2].

This chapter provides a primer on the use of Blockchain technology in social media. It begins with presenting an overview of Blockchain to make the chapter self-contained. It provides social media basics. It covers Blockchain social media and its applications. It highlights the benefits and challenges of Blockchain in social media. The last section concludes with comments.

11.2 OVERVIEW OF BLOCKCHAIN

Blockchain (BC) technology is a permanent record of online transactions. It is a distributed tamper-proof database, shared, and maintained by multiple parties. It is a new enabling technology that is expected to revolutionize many industries, including business. It has the potential for addressing significant business issues. The BC technology allows participants to move data in real-time, without exposing the channels to theft, forgery, and malice.

The term "Blockchain" refers to the way BC stores transaction data – in "blocks" that are linked together to form a "chain." The chain grows as the number of transactions increases. Since every entry is stored as a block on a chain, the care you receive is added to your personal ledger. The first Blockchain was conceived in 2008 by an anonymous person or group known as Satoshi Nakamoto, who published a white paper introducing the concept of a peer-to-peer electronic cash system he called Bitcoin [3]. Generic Blockchain images are shown in Figure 11.1 [4].

Figure 11.1   Generic Blockchain images [4].

At its core, Blockchain is a distributed system recording and storing transaction records. In a Blockchain system, there is no central authority. Instead, transaction records are stored and distributed across all network participants. Rather than having a centrally located database that manages records, the database is distributed to the networks and transactions are kept secure via cryptography. BC eliminates the need for a middleman that traditionally may facilitate such transactions. Figure 11.2 shows how Blockchain works [5].

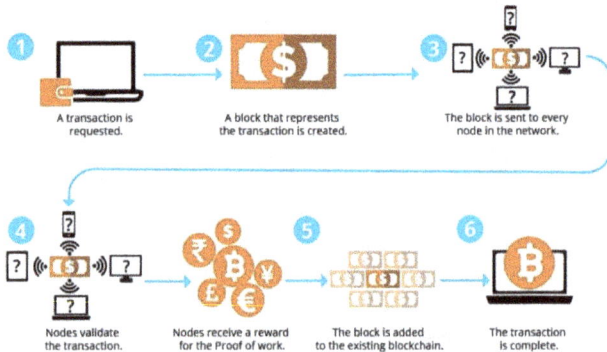

Figure 11.2 How Blockchain works [5].

Fundamentally, Blockchains are distributed digital database that record and maintain a list of transactions taking place in real time. They may also be regarded as decentralized ledgers that sequentially record transactions or interactions among users within a distributed network. They have the following properties [6]:

- Firstly, they are autonomous. They run on their own, without any person or company in charge.
- Secondly, they are permanent. They are like global computers with 100 percent uptime. Because the contents of the database are copied across thousands of computers, if 99 per cent of the computers running it were taken offline, the records would remain accessible and the network could rebuild itself.
- Thirdly, they are secure and tamper-proof. Each record in Blockchain is time stamped and stored cryptographically. The encryption used on Blockchains like Bitcoin and Ethereum is industry standard, open source, and has never been broken.

- Fourthly, they are open, allowing anyone to develop products and services on them.
- Fifthly, as Blockchain is a shared system, costs are also shared between all of its users.

The Blockchain was designed so transactions are immutable, i.e. they cannot be deleted. Thus, Blockchains are secure and meddle-free by design. Data can be distributed, but not copied. When it comes to digital assets and transactions, you can put almost anything on a Blockchain. Different scenarios call for different Blockchains. Blockchain is used in different areas such as depicted in Figure 11.3 [7].

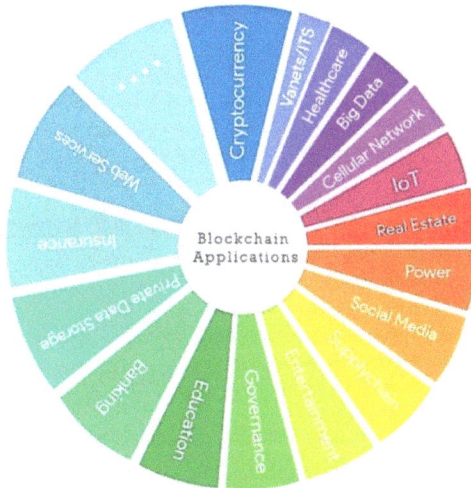

Figure 11.3 Applications of Blockchain [7].

The BC technology currently has the following features [8,9]:

1. *Peer-to-Peer (P2P) Network*: The first requirement of BC is a network, an infrastructure shared by multiple parties. This can be a LAN at a small scale or the Internet at a large scale. All nodes participating in a BC are connected in a decentralized P2P network. Transactions are broadcast to the P2P network. Due to some limitations of P2P networks, some vendors have provided cloud-based BCs.

2. *Cascaded Encryption*: A BC uses encryption to protect

transaction data. Blocks are encrypted in a cascaded manner, i.e. the encryption result of the previous block is used in encrypting the current block. The BC is secured by public key cryptography, with each peer generating its own public-private key pairs.

3. *Distributed Database*: A BC is digitally distributed across a number of computers. Each party on a BC has access to the entire database and no single party controls the data or the information. Since BC is decentralized, there is no need for central authorizes such as banks.

4. *Transparency with Pseudonymity*: Each node or participant on a Blockchain has a unique 30-plus-character alphanumeric address that identifies it. Users can choose to remain anonymous or provide proof of their identity to others.

5. *Irreversibility of Records*: Once a transaction is entered in the database and the accounts are updated, the records cannot be altered. Records on the database is permanent, chronologically ordered, and available to all others on the network.

There are two types of Blockchains: public and private. Public Blockchains are cryptocurrencies such as Bitcoin, enabling peer-to-peer transactions. Private Blockchains use Blockchain-based platforms such as Ethereum or Blockchain-as-a-service (BaaS) platforms running on private cloud infrastructure. A private BC is an intranet, while a public BC is the Internet. Companies will be disrupted the most by public Blockchains.

11.3 SOCIAL MEDIA BASICS

Social media (SM) is consumer-generated media that covers a variety of new sources of online information, created and used by consumers with the intent on sharing information with others. It employs mobile and web- based technologies to create, share, discuss, and modify consumer-generated content.

These are some common features of social media [10,11]:

1. *Accessibility*: They are easily accessible with little or no

cost.

2. *Connectedness*: They facilitate the development of online social networks by connecting people and bringing the world together.

3. *Communications*: They foster communication between individuals or organizations.

4. *Reach*: They offer unlimited reach to all content available to anyone, anywhere.

5. *News media*: They allow political news and information, true or not, to spread quickly.

6. *Collaboration*: They are computer-mediated technologies that facilitate the creation and sharing of information and ideas.

Social media takes on many forms. The six basic forms are [12]: (1) social networks such as Facebook and Twitter, (2) blogs - websites which allow users to subscribe, update, and leave comment, (3) wikis – collaborative website such Wikipedia which used to edit content, (4) podcasts – audio or video files that are published on the Internet, (5) content communities which share particular kinds of content, and (6) microblogging - allows instant publishing of content via Twitter. Blogs is probably the most commonly employed social media tool. Other forms include Internet forums, photographs or pictures, video, and social bookmarking. These and other activities on the social media are illustrated in Figure 11.4 [13].

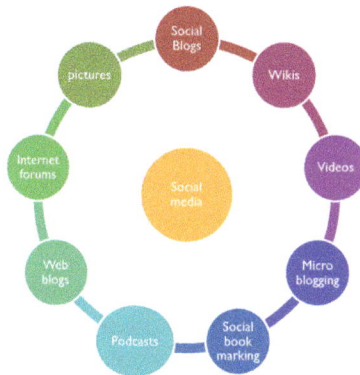

Figure 11.4 Activities on social media [13].

Social media is consumer-generated media that covers a variety of new sources of online information, created, and used by consumers with the intent on sharing information with others. It employs mobile and web-based technologies to create, share, discuss, and modify consumer-generated content. Consumers are most likely to leverage their power in social media to be more demanding of marketers [14]. The four most popular social media platforms are described here.

- *Facebook*: This is the most popular social media in the US and the rest of the world. It was launched on February 2004 by Mark Zuckerberg. Facebook can sensitize individuals (consumers) about many products and services. Different people use it to communicate with friends and family. A company can use Facebook to communicate their core values to a wide range of customers. Facebook consists of six primary components: personal profiles, status updates, networks (geographic regions, schools, companies), groups, applications, and fan pages.

- *Twitter*: Twitter was launched on July 2006 to provide a microblogging service. It allows individuals and companies to post short messages, share content, and have conversations with other Twitter users. Many Twitter posts (or "tweets") focus on the minutiae of everyday life.

- *LinkedIn*: This a networking website for the business community. It allows people to create professional profiles, post resumes, and communicate with other professionals. LinkedIn is where companies see the largest audiences.

- *YouTube*: YouTube has established itself as social media. It was launched in May 2005. It allows individuals to watch and share videos. YouTube may serve as home to aspiring filmmakers who might not have industry connections. YouTube can be both a blessing and a curse for some companies.

- *MySpace*: This social networking site bases its existence on advertisers who are paying for page views. It is an online community that allows you to meet your friends' friends, share photos, journals, and interests. It has a lot that users could do. There are MySpace sites in United Kingdom, Ireland, and Australia.

• *Instagram*: This is an image-based social media platform with more than 700 million active monthly users. The design is centered on a visual mobile experience. Instagram allows a simple and creative way to capture, edit, and share photos, videos and messages with friends and family.

• *TikTok*: This is the fastest growing social media platform of all time, with 800 users worldwide. This is a relatively new platform where users create and share short videos. Businesses are finding ways to use it as a marketing channel. They should tread lightly on TikTok, since most of its users are digital natives, who are very media savvy.

• *Discord*: This was introduced in 2015 and has total registered users of more than 390 million. Discord is initially used as a real-time messaging platform for the gamer community. Now it has a realm into the crypto community for its unique features like servers, channels, bots, etc.

Other social media include WhatsApp, Reddit, Pinterest, Flickr, Snapchat, WeChat, and Vine Camera. Some of these media are shown in Figure 11.5 [15].

Figure 11.5 Some social media [15].

Choosing the right social media platforms for your business is crucial. Social media allows you to do at least four important things [16]:

- Discover new ideas and trends.
- Connect with existing and new audiences in deeper ways.
- Bring attention and traffic to your work.
- Build, craft, and enhance your brand.

11.4 BLOCKCHAIN SOCIAL MEDIA

This is also known as Blockchain-enabled social media or decentralized social media. It is a social media platform that operates on a Blockchain. It is a decentralized ledger technology and therefore has no centralized authority. Instead of being a centralized hub that distributes information and data from the top down, it consists of individual blocks that are controlled by the user. Centralized systems surround and link to a single central component, while a distributed system is a collection of interconnected components with no central coordination. Figure 11.6 illustrates centralized, decentralized, and distributed networks [17].

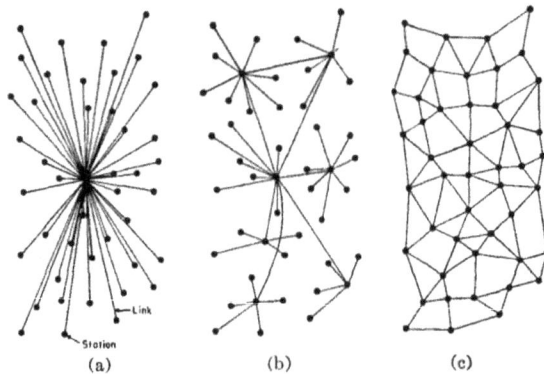

Figure 11.6   (a) Centralized, (b) decentralized, (c) distributed networks [17].

Despite having many advantages, a centralized architecture still contains several fundamental disadvantages [18]: (1) The content ownership is not in the hands of its creators; (2) Internet censorship is another thread of centralized architectures; (3) Due to their large userbase, social networks are being exploited for commercialization by marketing companies and advertising agencies. Due to these drawbacks, distributed social network architecture has been proposed as a reasonable substitution. Social networks that use a Blockchain are decentralized to varying degrees. Some use a Blockchain for data storage, some for monetization, and some for both data storage and monetization. Some of the Blockchain protocols that support the development of social media are Ethereum, Steem, and Stellar, to mention a few.

Many decentralized social networks exist as alternatives to established social media services, such as Facebook, LinkedIn, Twitter, and Medium. The decentralized social networks can fix many of the problems of traditional social networking platforms and improve users' overall experience. Decentralized social networks can be found on Bitcoin Cash, Ethereum, Steem, and numerous other Blockchain protocols. Compared to the currently popular centralized online social media, the Blockchain-based decentralized online social media platform presents a different business model. Social media networks built on the Blockchain are revolutionizing the way we share, interact and communicate with people online. Popular Blockchain-based social media sites include STEEM, Steemit, Verasity, Binded, Audius, Dtube, and Sapien. Blockchain-based social media platforms are taking social media on-chain. They platforms run the gamut from decentralized versions of Twitter, Medium, and LinkedIn to on-chain livestreaming and beyond.

There are three major concerns about the centralization of social media [19]:

- In social media networks, advertisers are the customers while users are the products. Social media platforms like Facebook earn revenues by selling users' personal data to targeted marketing or advertisers.
- Social media is also used as a propaganda platform for instigating users to think in a certain way regarding a social issue.
- Centralized services are easy to hack.

The benefits of decentralized social media (DeSo) include [20]:

- Decentralized social networks are censorship-resistant

and open to everyone. This means users cannot be banned, deplatformed, or restricted arbitrarily.

• Decentralized social networks are built on open-source ideals and make source code for applications available for public inspection. By eliminating the implementation of opaque algorithms common in traditional social media, Blockchain-based social networks can align the interests of users and platform creators.

• Decentralized social networks eliminate the "middle-man." Content creators have direct ownership over their content, and they engage directly with followers, fans, buyers, and other parties, with nothing but a smart contract in between.

• Decentralized social platforms offer an improved monetization framework for content creators via non-fungible tokens (NFTs), in-app crypto payments, and more.

• Decentralized social networks afford users a high level of privacy and anonymity. For instance, an individual can sign in to an Ethereum-based social network using an ENS profile or wallet—without having to share personally identifiable information (PII), such as names, email addresses, etc.

• Decentralized social networks rely on decentralized storage, not centralized databases, which are considerably better for safeguarding user data.

11.5 APPLICATIONS OF BLOCKCHAIN SOCIAL MEDIA

Blockchain technology has impact on digital marketing as well as social media. Here we consider some applications of blockchain-based social media.

1. *Decentralized Social Media*: This is already discussed in the previous section. Blockchains can help create decentralized social media networks that are user-controlled, censorship-resistant, and private. Blockchain social media are decentralized networking platforms built using Blockchain platforms that allow the development of applications and smart contracts. Decentralized social networks are content creation and distribution platforms running on the Blockchain. Decentralized social media platforms use decentralized Blockchain technology to store data across multiple nodes and ensure its integrity,

making it much more difficult to manipulate or censor. They are not under any central proprietary authority holding all the data.

2. *Blockchain Social Networks*: We now turn our attention to some of the popular social media networks powered by Blockchain. These include the following prominent Blockchain social media platforms [21]:

1. Peepeth
2. Sapien
3. Diaspora
4. Minds
5. All. me
6. Earn
7. SocialX
8. Steemit
9. Obsidian
10. Indorse

Some of these applications are shown in Figure 11.7 [7].

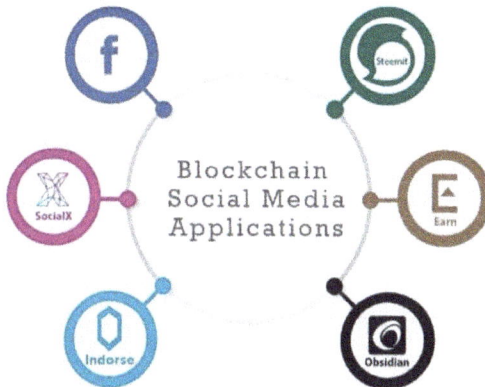

Figure 11.7   Blockchain social media applications [7].

3. *Digital Marketing*: Marketing is one of the areas that will get an enhancement from the use of Blockchain. The impact of Blockchain is significant is in the realm of digital marketing and advertising. Blockchain has allowed companies to target consumers with the correct marketing information and monitor their purchase history. Blockchain has allowed marketers to engage with social media users

to purchase details and other personal data [22].

4. *Crypto Social Media Marketing*: Marketing on social media platforms is the best way to reach a diverse audience. Every industry vertical including crypto and Blockchain projects starts to make use of social media to promote their product or service. Crypto social media marketing is a promotional activity or marketing strategy to market a crypto business, gain valuable insights, reach more potential investors, and optimize brand awareness. It can make use of various social media platforms like Facebook, Twitter, Instagram, LinkedIn, etc. to promote their brand. They are the best place to tell the world about a crypto project. The crypto community is a group of people on social media platforms that have discussions about crypto products or services. Crypto community marketing helps you to provide insights about your crypto projects to users and also get recommendations from them. Building trust among your audience is the key factor for crypto social media marketing [23].

5. *Digital Advertising*: Blockchain's decentralized nature provides transparency in digital advertising. It allows advertisers to verify the authenticity of ad impressions and clicks, ensuring that their budgets are spent on genuine interactions. This transparency builds trust between advertisers and publishers; it reduces the risk of ad fraud and improving overall campaign effectiveness. Blockchain's smart contract capabilities streamline payment processes in digital advertising. Smart contracts automatically execute transactions when predetermined conditions are met, eliminating the need for intermediaries and reducing transaction costs [24].

11.6 BENEFITS
Blockchain-based social media provide more benefits than just security and privacy. The "Blockchain + media" model is showing great vitality. Benefits of the Blockchain-based social media platforms include enhanced privacy, censorship-resistance, and the ability for users to both receive and send crypto via the social media platform itself. Because Blockchains are public and unalterable, transactions are fast, efficient and secure. Since Blockchain social media networks are decentralized, there is no central proprietary authority in charge of all data. Blockchain also provides encryption or validation techniques to protect user data. The Blockchain-based application puts the

interests of the users ahead of the corporation. Because there is no centralized authority, users on these networks have more privacy. Blockchain-based social media is fair in its distribution of revenue.

The primary benefits of Blockchain for social media include the following [25,26]:

- *Decentralization*: Blockchain technology allows for decentralized social media platforms where users control their data and have a stake in the network. The decentralization of social services is an opportunity to overcome the main privacy issues in social media, fake news, and censorship

This eliminates the need for a central authority or intermediary, promoting transparency and fairness.

- *Transparency*: The lack of transparency on social media sites is often criticized. Most sites have been charged with using arbitrary standards to determine what constitutes "acceptable" information. A Blockchain is intended to be completely transparent and what content is considered "acceptable" must be open, fair, and accurate in the application.

- *Immutability*: The Blockchain creates a distributed, immutable digital ledger, making it difficult to add or remove data covertly. Blockchain technology enables the establishment of social media platforms that guarantee immutability, i.e. making it impossible to alter or tamper with the information.

- *Confidentiality*: Users of decentralized social networks independently establish the rules for using the data generated by them, prohibiting and giving permission for their transfer, sale, and monetization. Anyone can be sure of their confidentiality, reliability, and honesty since the software is open source.

- *Privacy*: This is an essential aspect of social media. Corporations control these platforms and they can compromise user's privacy rights. Blockchain helps users in safeguarding their privacy. The Blockchain technology makes it possible to track data and monitor interaction with the content. Personal data remains in the system and never leaves it without the knowledge and agreement of the source. Users can manage their data and have full control over it. Blockchain-based social networks provide reliable protection for user data.

- *User Rights*: In centralized social networks, ordinary users have almost no rights. Services are provided "as is" and you cannot demand more quality, reliability or security. In decentralized systems, users manage the system by reaching a consensus on all common issues.

- *Lack of Censorship*: Popular Facebook, Instagram, Twitter or YouTube accounts often suffer because of allegedly violated site censorship rules. This is especially noticeable in countries where there are problems with freedom of speech. For example, in Russia, Iran, and China, accounts can be blocked. Political censorship is also present in America and Europe, but not as explicitly as in Russia, Iran and China. There are also problems with texts that talk about sex and violence, even if it is educational materials or scientific research. In decentralized social networks, the rules of censorship are set by users themselves.

- *Information Verification*: Fake information goes viral as fast as good content. However, the consequences of such events can be more dangerous than just a copyright violation. Blockchain helps fighting fake news due to its ledger system. Users can use such functions as data verification and personal verification. Everyone can verify content and IDs.

- *No Middlemen*: By using the Blockchain technology, business owners and digital marketers can talk directly to their audience, with no middlemen involved. Blockchain makes it possible to make secure consumer transactions right on a social media platform, with no need to use any third-party payment systems. Business owners get more accurate information from their audience, while consumers can choose what ads they want to see, making targeting much more effective.

11.7 CHALLENGES

There are rising concerns about the negative impact of social media. Just 27% of Americans say they have at least "some trust" in the information that comes from social media. Most Blockchains behave like expensive, slow databases that sacrifice efficiency for immutability and global consensus. Blockchains are logically centralized, exerting great effort to make many untrustworthy computers behave like one computer. We must be aware of the challenges facing Bockchaion-based social media networks, which include the following [27,28]:

- *Limitations*: Fundamental limitations of social media platforms include fake news, excessive trolling, censorship, data ownership, privacy, such as misinformation, lack of effective content moderation, digital piracy, data breaches, identity fraud, and demonetization.

- *Privacy Protection*: Although traditional social media's popularity is skyrocketing due to the enhanced marketing and entertainment opportunities it provides to its users, concerns about data and privacy breaches linked to these sites are growing. To join a typical social network, users must provide personally identifiable information (PII), such as their email address, home address, and phone number. These social platforms like Facebook and Twitter collect user data, which they then market to third parties for profit.

- *Regulations*: Regulations may be necessary to ensure that decentralized Blockchain social media's ideals are actualized to provide benefits and not harm.

- *Users Adoption*: Many people still view the idea of Blockchain technology with skepticism. Switching from centralized platforms will be challenging for the users given how addicted we are to some of the major social networks.

- *Market Instability*: It is a fact that the way the market functions in Web3 is unpredictable. As a result, a Blockchain-powered social network with built-in digital currencies may experience market volatility.

- *Social Media Tycoons*: There are obvious concerns about the major social media tycoons. They are frequently blamed for selling user data to advertising firms and have faced criticism for how they regulate content.

11.8 CONCLUSION

Blockchain technology is regarded one of the main disruptive technology of the millennium. Several research fields have tried to use it by exploiting its intrinsic characteristics. In many industries, Blockchain is considered as a breakthrough that can resolve data security concerns to a large extent. Several Web3 proponents believe they can establish a pioneer in the social media ecosystem with Blockchain that can overcome the centralized dominance of platforms.

Social networks, developed based on Blockchain technology, are becoming increasingly popular. Blockchain social media are decentralized networking platforms built using Blockchain protocols. They constitute a new layer-1 Blockchain created from the ground up to expand decentralized social applications to one billion users. Blockchain social media enables users to assert greater control over their data. Major social media companies are announcing plans to incorporate elements of decentralized social media.

Blockchain technology is a promising, yet not well understood, enabler of large-scale societal and economic change. Social media is evolving, and so is the technology that enables it. The relationship between social media and Blockchain will continue to evolve at lightning speed. More information about Blockchain in social media can be found in the books in [28-35].

REFERENCES

[1] H. Kodag et al., "DECENTRAGRAM: A Blockchain social media," International Journal of Scientific Development and Research, vol. 8, no. 4, April 2023, pp. 1649-1652.

[2] M. N. O. Sadiku, U. C. Chukwu, and J. O. Sadiku, "Blockchain in social media," American Journal of Social and Humanitarian Research, vol. 4, no. 9, September 2023, pp. 9-19.

[3] M. N. O. Sadiku, Y. Wang, S. Cui, and S. M. Musa, "A primer on Blockchain," International Journal of Advances in Scientific Research and Engineering, vol. 4, no. 2, February 2018, pp. 40-44.

[4] J. Graber, "Blockchain social networks: Using Blockchains for monetization and data storage," January 2020,

https://medium.com/decentralized-web/Blockchain-social-networks-c941fb337970

[5] P. di Torino, "Blockchain in finance," https://webthesis.biblio.polito.it/21440/1/tesi.pdf

[6] S. Depolo, "Why you should care about Blockchains: the non-financial uses of Blockchain technology," March 2016,

https://www.nesta.org.uk/blog/why-you-should-care-about-Blockchains-non-financial-uses-Blockchain-technology

[7] S. Johar, "Research and applied perspective to Blockchain technology: A comprehensive survey," Applied Sciences, vol.11, no. 14, July 2021.

[8] M. Iansiti and K. R. Lakhani, "The truth about Blockchain," Harvard Business Review, Jan./Feb. 2017.

https://hbr.org/2017/01/the-truth-about-Blockchain

[9] W. T. Tsai et al., "A system view of financial Blockchains," Proceedings of IEEE Symposium on Service-Oriented System Engineering, 2016, pp. 450-457.

[10] "Social media," Wikipedia, the free encyclopedia

https://en.wikipedia.org/wiki/Social_media

[11] V. Taprial and P. Kanwar, "Understanding social media,"

http://bookboon.com/en/understanding-social-media-ebook

[12] A. Mayfield, "What is Social Media? An e-book iCrossing,"
http://www.icrossing.com/uk/sites/default/files_uk/insight_pdf_files/What%20is%20Social%20Media_iCrossing_ebook.pdf

[13] S. Bowie, "Social work and the role of social media best practices,"
http://www.csus.edu/faculty/b/bowies/docs/what%20is%20social%20media%20use%20this.pdf

[14] C. Kohli, R. Surib, and A. Kapoor, "Will social media kill branding?" Business Horizons, 2015, vol. 58, pp. 35-44.

[15] A. Shaw, "How social media can move your business forward,"
https://www.forbes.com/sites/forbescommunicationscouncil/2018/05/11/how-social-media-can-move-your-business-forward/?sh=41d07cef4cf2

[16] S. Sreenivasan, "How to use social media in your career,"
https://www.nytimes.com/guides/business/social-media-for-career-and-business

[17] J. Graber, "Decentralized social networks," January 2020.
https://medium.com/decentralized-web/decentralized-social-networks-e5a7a2603f53

[18] H. H. Nguyen et al., "SoChainDB: A database for storing and retrieving Blockchain-powered social network data," SIGIR '22: Proceedings of the 45th International ACM SIGIR Conference on Research and Development in Information Retrieval, July 2022, pp. 3036–3045.

[19] "Decentralized social networks," https://ethereum.org/en/social-networks/

[20] L. Hertz, "Blockchain social media – Towards user-controlled data," https://www.leewayhertz.com/Blockchain-social-media-platforms/

[21] J. Bolander, "How Blockchain affects digital marketing & social media?" June 2021,
https://www.thedailymba.com/2021/06/21/how-Blockchain-affects-digital-marketing-social-media/

[22] "Top 6 powerful strategies for crypto social media marketing," November 2022,

https://shamlatech.com/crypto-social-media-marketing-strategies/

[23] Rittz Digital, "The impact of Blockchain technology on digital marketing and advertising,"

https://www.linkedin.com/pulse/impact-Blockchain-technology-digital-marketing-advertising

[24] Blockchain Council, "The race to build a social media platform on the Blockchain," January 2023,

https://www.Blockchain-council.org/Blockchain/the-race-to-build-a-social-media-platform-on-the-Blockchain/

[25] A. Silber, "Challenges of modern age: Blockchain can solve social media's problems," August 2023,

https://smallbusinessbonfire.com/Blockchain-social-media/

[26] "Blockchain social media – Towards user-controlled data,"

https://www.leewayhertz.com/Blockchain-social-media-platforms/#:~:text=Using%20the%20private%20key%2C%20the,only%20to%20the%20intended%20recipient.

[27] "Decentralized social networks 101,"

https://medium.com/klaytn/decentralized-social-networks-101-da65c19a599e

[28] A. Taylor, Cryptosocial: How Cryptocurrencies Are Changing Social Media. Business Expert Press, 2022.

[29] N. Upadhyay, Transforming Social Media Business Models Through Blockchain. Emerald Publishing Limited, 2019.

[30] G. Nash, Decentralized Social Media: The Future of Online Communication and Blockchain. Amazon Digital Services LLC, 2023.

[31] T. Lamonica, Peer-to-Peer Decentralized Layer Blockchain Technology. University of Ottawa, 2020.

[32] M. Johnsen, Blockchain in Digital Marketing: A New Paradigm of Trust. Independently Published, 2020.

[33] G. Destefanis and M. Ragnedda (eds.), Blockchain and Web 3.0: Social, Economic, and Technological Challenges. Taylor &

Francis,2019.

[34] B. Nagpa, N. Saxena, R. S. Bhadoria (eds.), Blockchain Technology for Secure Social Media Computing. Institution of Engineering & Technology, 2023.

[35] D. Le, G. Shrivastava, and K. Sharma, Cryptocurrencies and Blockchain Technology Applications. Wiley, 2020.

CHAPTER 12

BLOCKCHAIN IN ARTIFICIAL INTELLIGENCE

"Some people call this artificial intelligence, but the reality is this technology will enhance us. So instead of artificial intelligence, I think we'll augment our intelligence."

— Ginni Rometty

12.1 INTRODUCTION

A Blockchain is a distributed, decentralized, immutable ledger used to store encrypted data. On the other hand, Artificial intelligence (AI) is the engine or the "brain" that will enable analytics and decision making from the data collected. It is a technology that can perform complex tasks that require human intelligence, and it holds the potential of exceeding human capabilities. The amalgamation of AI and Blockchain holds great potential to create new business models enabled through digitalization. The combination of AI and Blockchain can take place in multiple dimensions. There are examples where AI and Blockchain can be integrated for process improvement and value creation (e.g., asset management, customer service, dispute resolution, fraud prevention, production evaluation, supply chain monitoring) [1].

Blockchain and artificial intelligence (AI) are important disruptive, emerging technologies. When both of them are combined, they can help you build an immutable, safe, and decentralized system. This will lead to major data and information security advances in different industries. We have witnessed that AI and Blockchain are the two leading technologies worldwide, and they have attracted wide attention in both academia and industry. Their collaboration has the potential to push data exploitation to newer heights and has brought many new opportunities. For example, their joint application has helped the healthcare industry to navigate the COVID-19 crisis [2].

This chapter considers the combination of AI and Blockchain and its applications. It begins with providing an overview on Blockchain to make the chapter self-contained. It covers the fundamentals of AI. It presents Blockchain AI. It presents the applications of combining

AI and Blockchain. It highlights the benefits and challenges of the integration of Blockchain and AI. The last section concludes with comments.

12.2 OVERVIEW OF BLOCKCHAIN

Blockchain (BC) technology is a permanent record of online transactions. It is a distributed tamper-proof database, shared, and maintained by multiple parties. It is a new enabling technology that is expected to revolutionize many industries, including business. It has the potential for addressing significant business issues. The BC technology allows participants to move data in real-time, without exposing the channels to theft, forgery, and malice.

The term "Blockchain" refers to the way BC stores transaction data – in "blocks" that are linked together to form a "chain." The chain grows as the number of transactions increases. Since every entry is stored as a block on a chain, the care you receive is added to your personal ledger. The first Blockchain was conceived in 2008 by an anonymous person or group known as Satoshi Nakamoto, who published a white paper introducing the concept of a peer-to-peer electronic cash system he called Bitcoin [3]. A typical Blockchain architecture is shown in Figure 12.1 [4].

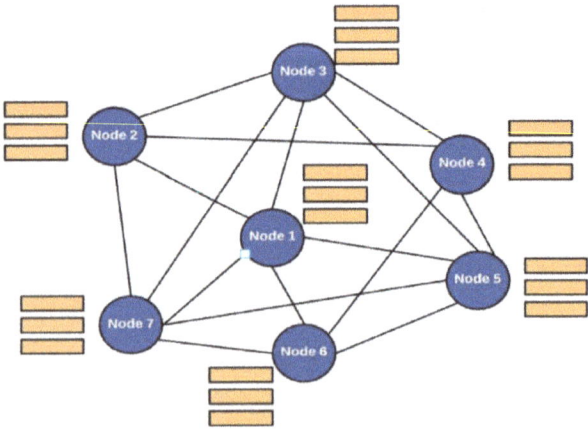

Figure 12.1 A Blockchain architecture [4].

At its core, Blockchain is a distributed system recording and storing transaction records. In a Blockchain system, there is no central authority. Instead, transaction records are stored and distributed across all network participants. Rather than having a centrally located database that manages records, the database is distributed to the networks and transactions are kept secure via cryptography. BC eliminates the need for a middleman that traditionally may facilitate such transactions. Figure 12.2 shows how Blockchain works [5].

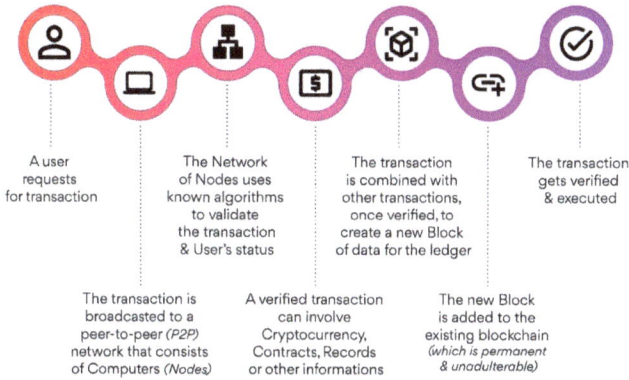

Figure 12.2   How the Blockchain technology works [5].

Fundamentally, Blockchains are distributed digital database that record and maintain a list of transactions taking place in real time. They may also be regarded as decentralized ledgers that sequentially record transactions or interactions among users within a distributed network. They have the following properties [6]:

- Firstly, they are autonomous. They run on their own, without any person or company in charge.

- Secondly, they are permanent. They are like global computers with 100 percent uptime. Because the contents of the database are copied across thousands of computers, if 99 per cent of the computers running it were taken offline, the records would remain accessible and the network could rebuild itself.

- Thirdly, they are secure and tamper-proof. Each record in Blockchain is time stamped and stored cryptographically. The encryption used on Blockchains like Bitcoin and Ethereum is industry standard, open source, and has never been broken.

- Fourthly, they are open, allowing anyone to develop products and services on them.

- Fifthly, as Blockchain is a shared system, costs are also shared between all of its users.

The Blockchain was designed so transactions are immutable, i.e. they cannot be deleted. Thus, Blockchains are secure and meddle-free by design. Data can be distributed, but not copied. When it comes to digital assets and transactions, you can put almost anything on a Blockchain. Different scenarios call for different Blockchains. Blockchain is used in different areas such as depicted in Figure 12.3 [7].

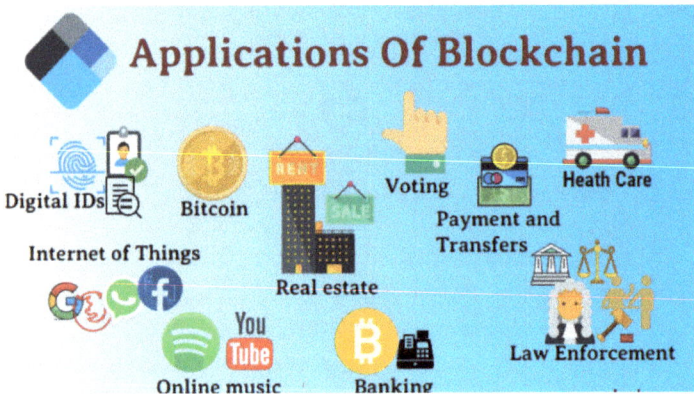

Figure 12.3   Different uses of Blockchain [7].

The BC technology currently has the following features [8,9]:

1. *Peer-to-Peer (P2P) Network*: The first requirement of BC is a network, an infrastructure shared by multiple parties. This can be a LAN at a small scale or the Internet at a large scale. All nodes participating in a BC are connected in a decentralized P2P network. Transactions are broadcast to the P2P network. Due to some limitations of P2P networks, some vendors have provided cloud-based BCs.

2. *Cascaded Encryption*: A BC uses encryption to protect transaction data. Blocks are encrypted in a cascaded manner, i.e. the encryption result of the previous block is used in encrypting the current block. The BC is secured by public

key cryptography, with each peer generating its own public-private key pairs.

3. *Distributed Database*: A BC is digitally distributed across a number of computers. Each party on a BC has access to the entire database and no single party controls the data or the information. Since BC is decentralized, there is no need for central authorizes such as banks.

4. *Transparency with Pseudonymity*: Each node or participant on a Blockchain has a unique 30-plus-character alphanumeric address that identifies it. Users can choose to remain anonymous or provide proof of their identity to others.

5. *Irreversibility of Records*: Once a transaction is entered in the database and the accounts are updated, the records cannot be altered. Records on the database is permanent, chronologically ordered, and available to all others on the network.

There are two types of Blockchains: public and private. Public Blockchains are cryptocurrencies such as Bitcoin, enabling peer-to-peer transactions. Private Blockchains use Blockchain-based platforms such as Ethereum or Blockchain-as-a-service (BaaS) platforms running on private cloud infrastructure. A private BC is an intranet, while a public BC is the Internet. Companies will be disrupted the most by public Blockchains.

12.3 FUNDAMENTALS OF ARTIFICIAL INTELLIGENCE

Artificial intelligence (AI) is one of the most important global issues of the 21st century. The term "artificial intelligence" (AI) was coined in 1956 by John McCarthy during a conference held on this subject. AI is the branch of computer science that deals with designing intelligent computer systems that mimic human intelligence. The ability of machines to process natural language, to learn, to plan makes it possible for new tasks to be performed by intelligent systems. The main purpose of AI is to mimic the cognitive function of human beings and perform activities that would typically be performed by a human being. AI is stand-alone independent electronic entity that functions much like human healthcare expert. Today, AI is integrated into our daily lives in several forms, such as personal assistants, automated mass transportation, aviation, computer gaming, facial recognition at passport control, voice recognition on virtual assistants, driverless cars, companion robots, etc. AI technologies are performing better and better at analyzing data [10,12].

An important feature of AI technology is that is can be added to existing technologies. AI has benefited many areas such chemistry and medicine, where routine diagnoses can initiated by AI-aided computers. It embraces a wide range of disciplines such as computer science, engineering, chemistry, biology, physics, astronomy, neuroscience, and social sciences.

AI is not a single technology but a range of computational models and algorithms. The major disciplines in AI include expert systems, fuzzy logic, and artificial neural networks (ANNs), machine learning, deep learning, natural language processing, computer vision, and robotics. The various computer-based tools or technologies that have been used to achieve AI's goals are the following [12,13]:

- *Expert Systems*: An expert system (ES) (or knowledge-based system) enables computers to make decisions by interpreting data and selecting between alternatives just as a human expert would do. It uses a technique known as rule-based inference in which rules are used to process data.

- *Neural Networks*: These computer programs identify objects or recognize patterns after having been trained. Artificial neural networks (ANNs) are parallel distributed systems consisting of processing units (neurons) that calculate some mathematical functions. The ANN model represents nonlinear relationships which are directly learned from the data being modeled. Neural networks are being explored for healthcare applications in imaging and diagnoses, risk analysis, lifestyle management and monitoring, health information management, and virtual health assistance.

- *Natural Language Processors*: Computer programs that translate or interpret language as it is spoken by normal people. NLP techniques extract information from unstructured data such as clinical notes to supplement and enrich structured medical data. NLP includes applications such as speech recognition, text analysis, translation and other goals related to language. There are two basic approaches to NLP: statistical and semantic. Healthcare is the biggest user of the NLP tools [14].

• *Robots*: Computer-based programmable machines that have physical manipulators and sensors. The introduction of intelligent robots in the healthcare domain enhances patients' satisfaction, accuracy of diagnosis, and operational efficiency of hospitals. Medical robots can help with surgical operations, rehabilitation, social interaction, assisted living, etc. Robotic-guidance is becoming common in spine surgery [15].

• *Fuzzy Logic*: Reasoning based on imprecise or incomplete information in terms of a range of values rather than point estimates. Fuzzy logic deals with uncertainty in knowledge that simulates human reasoning in incomplete or fuzzy data. The fuzzy model is robust to parameter changes and tolerant to impression.

• *Machine Learning*: Algorithms to make predictions and interpret data and "learn", without static program instructions. ML is a statistical technique for fitting models to data and training models with data. ML extracts features from input data by constructing analytical data algorithms and examines the features to create predictive models. The most common ML algorithms are supervised learning, unsupervised learning, reinforcement learning, and deep learning. The most common application of ML is precision medicine. ML algorithms are a good fit for anti-malware solutions because machine learning is well suited to solve 'fuzzy' problems.

• *Deep Learning*: A subset of machine learning built on a deep hierarchy of layers, with each layer solving different pieces of a complex problem. It aims at increasing the capacity of supervised and unsupervised learning algorithms for solving complex real-world problems by adding multiple processing layers.

• *Data Mining*: This deals with the discovery of hidden patterns and new knowledge from large databases. Data mining exhibits a variety of algorithmic tools such as statistics, regression models, neural networks, fuzzy sets, and evolutionary models.

Some of these AI tools are illustrated in Figure 12.4 [5].

Figure 12.4   *AI tools or branches [5].*

Each AI tool has its own advantages. Using a combination of these models, rather than a single model, is recommended. AI technologies are drastically influencing the retail industry and customer experience. Applications of AI technologies for cybersecurity tasks are attracting greater attention from the private and the public sectors due to the rate at which threats are developing.

12.4 BLOCKCHAIN AI

Blockchain is an expensive medium of storing vast data in a traditional method. It is popular for its decentralization and transparency. Therefore, it gives the perfect instrument for peeling the layers of complex AI algorithms to understand their decision-making processes. Blockchain is a shared and permanent ledger that is being used for the encryption of data, while AI enables an individual to analyze and make decisions from the collected data. Combining the two technologies will have multiple complexions but will provide many benefits. AI approaches the use of Blockchain for providing decentralized learning by facilitating the secure sharing of knowledge and trust in the decision-making process.

Blockchain technology is interlinked with AI in many ways. Below are the major integrations [16]:

- *Authenticity*: Blockchain technology can help validate the authenticity of images, video files, text documents, or other types of media by being able to cryptographically verify where a piece of content originates from and whether it has been tampered with or altered in any way. The integration of AI and Blockchain can help to ensure the authenticity of information about credit evaluation and the verification of ensuing financial transactions by traders.

- *Augmentation*: AI understands and processes the data at a great speed by bringing higher intelligence to the Blockchain-based business networks. By providing access to large volumes of data from within and outside of the organization, Blockchain helps AI scale to provide more actionable insights, manage data usage and model sharing, and create a trustworthy and transparent data economy.

- *Automation*: Blockchain technology ensures that the automation process is quicker than centralized databases. It will also ensure that only admins will be able to modify the data that is sent to the machine learning model. Automation, AI, and Blockchain will bring newer values to business processes that span different parties like adding, removing friction, and increasing speed and efficiency. The integration of AI and Blockchain facilitates the automation business.

Figure 12.5 compares the properties of AI and Blockchain [17].

Blockchain	AI
Immutable	Modeling and adjusting throughout time
Deterministic	Probabilistic
Can be described to human users and is transparent since it can be monitored.	As judgments are determined by machine learning systems, they cannot be communicated to human users and are hence opaque.
Infrastructure that is decentralized and distributed	Powered by a central infrastructure

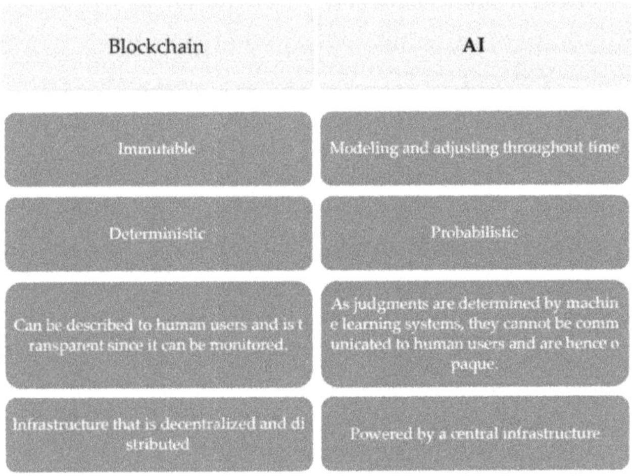

Figure 12.5   Properties of AI and Blockchain [17].

12.5 APPLICATIONS OF BLOCKCHAIN AND AI

AI and Blockchain are the key technologies propelling the wave of digital transformation. By combining the powerful analytical capabilities of AI with the secure, decentralized nature of Blockchains, the technologies could be applied to wide range areas such as education, healthcare, energy, social impact, agriculture, urban planning, data-driven decision-making, autonomous vehicles, finance, smart cities, and 6G networks. We will consider some of these applications [18,19]:

- *Smart Computing Power*: Operating a Blockchain requires large processing power. AI affords us the opportunity to tackle tasks in a more intelligent and efficient way.

- *Creating Diverse Data Sets*: Unlike artificial intelligence based-projects, Blockchain technology creates decentralized, transparent networks that can be accessed by anyone, around the world. While Blockchain technology is the ledger that powers cryptocurrencies, Blockchain networks are now being applied to a number of industries to create decentralization. Diverse algorithms can be built on diverse data sets.

- *Data Protection*: The progress of AI is completely

dependent on the input of data. Basically, data feeds AI, and through it, AI will be able to continuously improve itself. On the other hand, Blockchain is essentially a technology that allows for the encrypted storage of data on a distributed ledger. When combining Blockchains with AI, we have a backup system for the sensitive and highly valuable personal data of individuals.

• *Data Monetization*: Another disruptive innovation that could be possible by combining the two technologies is the monetization of data. Monetizing collected data is a huge revenue source for large companies, such as Facebook and Google.

• *Trusting AI Decision Making*: As AI algorithms become smarter through learning, it will become increasingly difficult for data scientists to understand how these programs came to specific conclusions and decisions. Through the use of Blockchain technology, there are immutable records of all the data, variables, and processes used by AIs for their decision-making processes.

Other applications include the following [19]:

• *Healthcare*: The healthcare industry is another sector where the convergence of AI and Blockchain has enormous potential. Blockchain protects privacy and increases security of health data and enables the secure storage of patient data. When access is granted, health professionals gain insights from this data through the use of AI. Integration of AI and Blockchain create predictive system contributing to clinical workflow. Combining advanced data analysis with a decentralized framework for clinical trials enables data integrity, transparency, patient tracking, consent management and automation of trial participation and data collection.

• *Financial Services*: Blockchain and AI are set to transform financial markets. The convergence of the two technologies is also re-inventing the financial services industry by increasing the speed of transactions and enabling trust among transacting parties. The amalgamation of the two technologies has also introduced decentralized autonomous business models that brings greater flexibility, agility, and cost-effectiveness to business. Blockchain and AI are transforming the financial services industry by enabling trust, removing friction from multiparty transactions, and accelerating the speed of transactions.

- *Supply Chain*: AI can help enhance the efficiency and transparency of supply chains in the transportation sector and other industries. By digitizing a largely paper-based process, making the data shareable and trustworthy, and adding intelligence and automation to execute transactions, AI and Blockchain are transforming supply chains across industries and creating new opportunities.

- *Security*: Decentralized, Blockchain-based systems have been designed from the ground up to combat manipulation by various adversaries, and these security measures could extend to the use of adversarial AI agents. With the implementation of AI, Blockchain technology becomes safer by making secure future application deployments. The utility of AI models and the security of Blockchains can help reduce attack vectors and bolster the security of AI applications.

- *Education*: The future of education relies on advances in and use of smart technologies, particularly those involving Blockchain and artificial intelligence.

- *Smart City*: A smart city is a technologically advanced urban area where intelligent subsystems connect people and organizations. It uses large data sets to offer stakeholders real-time access to high-quality public services and thus improve the quality of life in the city. AI and Blockchain have great potential to support the development of smart city. AI and Blockchain can provide significant benefits to many areas of the city's functioning: it is a huge database for collecting and analyzing data. Blockchain consensus methods allow greater transparency and less susceptibility to manipulation. AI and Blockchain support smart city as depicted in Figure 12.6 [20].

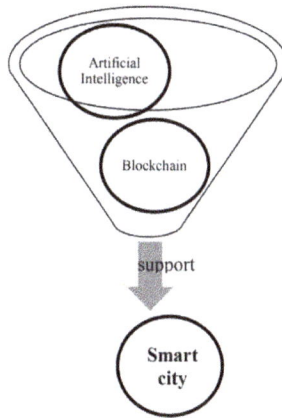

Figure 12.6   AI and Blockchain support smart city [20].

Some of these applications of Blockchain and AI combined are illustrated in Figure 12.7 [21].

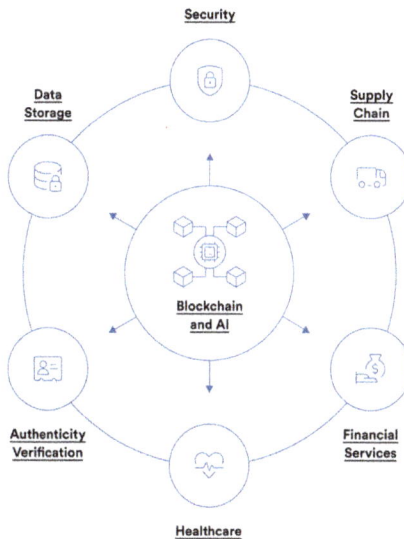

Figure 12.7   Applications of Blockchain and AI combined [21].

231

12.6 BENEFITS

Across industries, combining AI with Blockchain delivers new opportunities and brings new value to business. The convergence of AI and Blockchain has the potential to provide numerous benefits beyond traditional business applications. It could unlock entirely new business models, create operational efficiencies for organizations, help automate repetitive tasks for individuals, enable more secure and efficient data exchange, enhance decision-making processes through AI-driven smart contracts, and improve overall trust and transparency in key infrastructure and economic processes. AI and Blockchain are proving to be quite a powerful combination, improving virtually every industry in which they are implemented. The confluence of AI in Blockchain creates perhaps what is the world's most reliable technology-enabled decision-making system that is virtually tamper-proof and provides solid insights and decisions. Other benefits of the confluence of AI in Blockchain include the following [22,23]:

- *Security*: Blockchain offers a secure way to encrypt sensitive information With the implementation of AI, Blockchain technology becomes safer by making secure future application deployments. AI and Blockchain could provide a substantial boost for improvements in encryption. The inherent encryption in a Blockchain ensures that data is well-protected.

- *Efficiency*: AI can introduce even new decentralized learning systems such as federated learning or new data-sharing techniques that make the system much more efficient. Blockchains are suitable for storing sensitive personal data that may offer value and convenience when intelligently handled with AI.

- *Trust*: When Blockchain is applied in conjunction with AI, users have clear records to follow the system's thinking process. This, in turn, helps the bots trust each other, increasing machine-to-machine interaction and allowing them to share data and coordinate decisions at large.

- *Better Management*: The benefit of better management is an obvious reason for which it is important to consider AI and Blockchain combinations. When it comes to cracking codes, human experts get better over time with practice. So, AI additionally helps in managing Blockchain systems better.

• *Privacy*: Privacy protection techniques arising from the integration of AI and Blockchain are of notable significance. Making private data secure invariably leads to it being sold, resulting in data markets/model markets. Blockchain facilitates AI applications in secure data sharing, preserving data privacy, and supporting trusted AI decision.

• *Efficient Storage*: Blockchains are ideal for storing the highly sensitive, personal data which, when smartly processed with AI, can add value and convenience.

• *Automation*: AI can help enhance the performance of Blockchain networks by automating processes and improving accuracy. By automating processes and analyzing data on the Blockchain, financial institutions can improve their risk management and compliance processes.

• *Financial Services*: As mentioned earlier, one of the most significant use cases of AI and Blockchain is in the financial industry. Blockchain and AI are transforming the financial services industry before our eyes. Financial institutions deal with vast amounts of data. AI and Blockchain together can help manage this data more efficiently. The combination is poised to transform the accounting profession.

Some of these benefits are illustrated in Figure 12.8 [24].

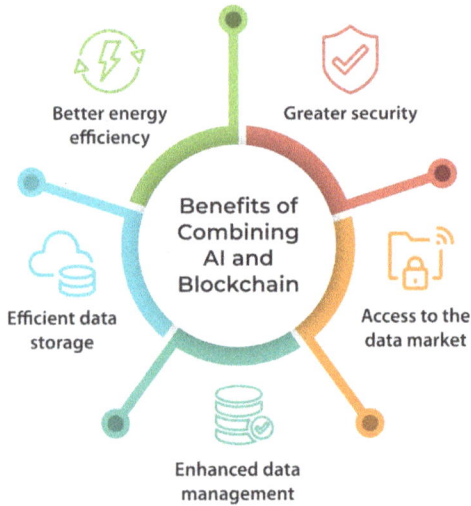

Figure 12.8   Some of the benefits of combining AI and Blockchain [24].

12.7 CHALLENGES

Integrating Blockchain and AI is not a straightforward task, as they have different characteristics, requirements, and challenges. While Blockchain struggles with scalability and efficiency, AI struggles with transparency and privacy, which makes the two technologies the perfect match because each can address the other's weaknesses. Blockchain provides the trust, privacy, and accountability to AI, while AI provides the scalability, efficiency, and security. Blockchain's immutable digital records may be a way to offer insights into AI's framework and model to address the challenge of transparency and data integrity. While Blockchain and AI definitely make a good pair, there are many challenges and obstacles in their way. These challenges include [25-28]:

- *Compatibility*: Making AI and Blockchain work together requires ensuring compatibility factors. AI relies on large amounts of data, which can be stored and processed in centralized servers or cloud platforms. On the other hand, Blockchain is a decentralized network of nodes that store and verify transactions in encrypted blocks. To make them work together, we need to find a way to exchange data between the centralized and decentralized environments, without compromising security, privacy, or performance. Solutions like hybrid architectures, smart contracts, oracles, and off-chain computation can bridge the gap by facilitating secure data exchange between centralized and decentralized environments. This ensures compatibility while maintaining security, privacy, and performance, allowing the benefits of both AI and Blockchain to be harnessed effectively.

- *Scalability*: This is a critical technical roadblock when integrating AI and Blockchain technologies due to varying requirements, parameters, and limitations, such as processing speed, data handling, and resource consumption. To overcome these issues, we need to optimize the design and configuration of AI and Blockchain systems. Integrating AI and Blockchain presents challenges due to their differing characteristics. Scalability is a common issue, with AI needing high-speed processing and Blockchain having limited throughput. To address this, optimization and techniques like sharding are essential. Ensuring data privacy and security is crucial, as Blockchain's transparency may conflict with AI's need for private data; this can be managed through encryption and permissioned Blockchains.

- *Governance*: Integrating AI and Blockchain faces governance issues. AI and Blockchain have different models of governance, which can affect how they are regulated, controlled, and audited. AI is often governed by centralized authorities, while Blockchain is governed by decentralized protocols. To harmonize the two technologies, you need to establish clear and consistent rules and standards for your AI and Blockchain systems.

- *Ethical issues*: Another challenge of integrating AI and Blockchain concerns ethical issues. AI and Blockchain have both positive and negative impacts on society, economy, and environment, which can raise ethical questions and dilemmas. To address these issues, you need to adopt ethical principles and frameworks for your AI and Blockchain systems.

- *Skills gap*: A significant challenge that arises at the

intersection of Blockchain and artificial intelligence is skill gap. AI and Blockchain are complex and evolving technologies, which require specialized knowledge and skills to develop, deploy, and maintain. However, there is a shortage of qualified and experienced professionals who can master both domains. To bridge this gap, you need to invest in education and training programs for your AI and Blockchain teams.

• *Testing and Debugging*: A major challenge of integrating AI and Blockchain is testing and debugging. Testing and debugging them is not easy, as they have different methods, tools, and standards. AI systems are often tested and debugged using data-driven approaches, such as validation, verification, or simulation. Blockchain systems are often tested and debugged using code-driven approaches, such as formal methods, audits, or reviews. To test and debug your AI and Blockchain systems, you need to use a combination of these approaches.

• *Data Privacy and Security*: Another challenge at the intersection of Blockchain and artificial intelligence is data privacy and security. When AI algorithms are integrated with Blockchain systems, sensitive data may be exposed to all participants in the network. Maintaining data privacy while leveraging the benefits of decentralized AI is a complex issue.

• *Computation*: A core challenge at the crossroads of Blockchain and AI is computation requirement. Both Blockchain and AI have demanding computational requirements. When you combine both, you end up with an even greater strain. This can lead to bottlenecks and delays.

• *Legal & Regulatory Implications*: Both AI and Blockchain need regulation to address risks. Malicious actors can exploit unregulated, decentralized systems. Data privacy and protection are the primary concerns when exposing sensitive data regulated by a Blockchain to AI models. Regulation policies strictly force businesses to handle client data by ensuring consensual usage of data and information. The legal issues related to smart contracts are challenging. Therefore, it is mandatory to create contractual terms and conditions carefully.

- *Public Perception*: Misconceptions persist regarding both technologies. For example, some erroneously equate Blockchain with cryptocurrencies, overlooking its broader applications. AI myths, including the fear of machines taking over the world, create uncertainty. Some perceive Blockchain and AI as oil and water that do not mix

- *Environmental Sustainability*: Both AI and Blockchain come with environmental concerns related to energy consumption. As evidenced by models like ChatGPT, AI can require substantial operations power. Running Blockchain systems, especially for cryptocurrencies like Bitcoin, demands significant energy resources.

These challenges can limit the practicality and efficiency of AI applications on Blockchain. Overcoming them requires innovative solutions and proactive efforts from tech companies. In spite of all the development and advancements taking place in the Blockchain and AI space, there is still a long road ahead of them.

12.8 CONCLUSION

Artificial intelligence and Blockchain are two of the most transformative and disruptive technologies of our times. They are the most cutting-edge technologies and will reshape how people work, interact, and live. Blockchain is a distributed immutable ledger system that can be used for implementing cryptocurrency with security, while AI enables machines to make intelligent decisions and develop self-learning capabilities. The combination of the two technologies has gained a lot of attention in recent years, especially since such integration can improve security, efficiency, and productivity of applications in business environments characterized by volatility, uncertainty, complexity, and ambiguity [29].

The idea of fusing Blockchain technology and artificial intelligence is still in its early stages. The combination can improve current business practices and introduce new business models that can act as independent economic agents making decisions autonomously. The blend of AI and Blockchain could introduce radical innovations in the future. The challenges for combining both play a major role in defining their future together. AI and Blockchain are promising technologies for all nations. More information about Blockchain in AI can be found in the books in [30-47].

REFERENCES

[1] "Artificial intelligence and Blockchain integration in business: Trends from a bibliometric-content analysis," April 2022,

https://link.springer.com/article/10.1007/s10796-022-10279-0

[2] M. N. O. Sadiku, U. C. Chukwu, and J. O. Sadiku, "Blockchain in artificial intelligence," Horizon: Journal of Humanities and Artificial Intelligence, vol. 2, no. 9, 2023, pp. 7-17.

[3] M. N. O. Sadiku, Y. Wang, S. Cui, and S. M. Musa, "A primer on Blockchain," International Journal of Advances in Scientific Research and Engineering, vol. 4, no. 2, February 2018, pp. 40-44.

[4] M. J. Tuyisenge, "Blockchain technology security concerns: Literature review,"

https://www.diva-portal.org/smash/get/diva2:1571072/FULLTEXT01.pdf

[5] "The future of AI and Blockchain technology & how it complements each other?

https://www.turing.com/kb/how-Blockchain-and-ai-complement-each-other

[6] P. Pedamkar, "Applications of Blockchain," June 2023,

https://www.educba.com/applications-of-blockchain/

[7] O. Bheda, "What is Blockchain?" https://builtin.com/Blockchain

[8] M. Iansiti and K. R. Lakhani, "The truth about Blockchain," Harvard Business Review, Jan./Feb. 2017.

https://hbr.org/2017/01/the-truth-about-Blockchain

[9] W. T. Tsai et al., "A system view of financial Blockchains," Proceedings of IEEE Symposium on Service-Oriented System Engineering, 2016, pp. 450-457.

[10] M. N. O. Sadiku, T. J. Ashaolu, and S. M. Musa, "Artificial intelligence in medicine: A primer," International Journal of Trend in Research and Development, vol. 6, no. 1, Jan.-Feb. 2019, pp. 270-272.

[11] Y. Mintz and R. Brodie, "Introduction to artificial intelligence in medicine," Minimally Invasive Therapy & Allied Technologies,

vol. 28, no. 2, 2019, pp. 73-81.

[12] R. O. Mason, "Ethical issues in artificial intelligence," Encyclopedia of Information Systems, vol 2, 2003, pp. 239-258.

[13] A. N. Rames et al., "Artificial intelligence in medicine," Annals of the Royal College of Surgeons of England, vol. 86, 2004, pp. 334–338.

[14] M. N. O. Sadiku, Y. Zhou, and S. M. Musa, "Natural language processing in healthcare," International Journal of Advanced Research in Computer Science and Software Engineering, vol. 8, no. 5, May 2018, pp. 39-42.

[15] "No longer science fiction, AI and robotics are transforming healthcare,"

https://www.pwc.com/gx/en/industries/healthcare/publications/ai-robotics-new-health/transforming-healthcare.html

[16] "The future of AI and Blockchain technology & how it complements each other?"

https://www.turing.com/kb/how-Blockchain-and-ai-complement-each-other

[17] H. Taherdoost, Blockchain Technology and Artificial Intelligence Together: A Critical Review on Applications," Applied Sciences, vol. 12, December 2022.

[18] A. Banafa, "Blockchain and AI: A perfect match?"

https://www.bbvaopenmind.com/en/technology/artificial-intelligence/Blockchain-and-ai-a-perfect-match/

[19] "Blockchain and artificial intelligence (AI),"

https://www.ibm.com/topics/Blockchain-ai

[20] S. Kauf, "Artificial intelligence and Blockchain for smart city,"

https://bibliotekanauki.pl/articles/2204746

[21] "Use cases of AI in Blockchain," May 2023,

https://blog.chain.link/Blockchain-ai-use-cases/

[22] "Integration of Blockchain and AI,"

https://www.geeksforgeeks.org/integration-of-Blockchain-and-ai/

[23] J. Murphy, "Use cases show the combined potential of AI and Blockchain,"

https://www.techtarget.com/searchenterpriseai/tip/Use-cases-show-the-combined-potential-of-AI-and-Blockchain#:~:text=AI%20can%20help%20enhance%20the%20efficiency%20and%20transparency%20of%20supply,companies%20to%20optimize%20their%20operations.

[24] "Does artificial intelligence impact Blockchain technology?"

https://www.turing.com/kb/does-artificial-intelligence-impact-Blockchain-technology

[25] "How can you overcome AI and Blockchain integration challenges?" October 2023,

https://www.linkedin.com/advice/0/how-can-you-overcome-ai-Blockchain-integration#:~:text=To%20overcome%20these%20issues%2C%20you,increase%20efficiency%20and%20reduce%20costs.&text=Integrating%20AI%20and%20Blockchain%20presents%20challenges%20due%20to%20their%20differing%20characteristics.

[26] H. Sajid, "Exploring the intersection of AI and Blockchain: Opportunities & challenges,"

https://www.unite.ai/exploring-the-intersection-of-ai-and-Blockchain-opportunities-challenges/

[27] "What are some challenges in the intersection of Blockchain and artificial intelligence?" August 2023,

https://blocktelegraph.io/Blockchain-and-artificial-intelligence/

[28] "AI vs Blockchain: Which will go mainstream first?" August 2023,

https://www.analyticsinsight.net/ai-vs-Blockchain-which-will-go-mainstream-first/#:~:text=AI%20Leads%20Mass%20Adoption%20while,Blockchain%20intensifies%20for%20mainstream%20adoption.

[29] V. Charles, A. Emrouznejad, and T. Gherman, "A critical analysis of the integration of Blockchain and artificial intelligence for supply chain," Annals of Operations Research, vol. 327, January 2023, pp. 7–47

[30] S. S. Smith, Blockchain, Artificial Intelligence and Financial Services: Implications and Applications for Finance and Accounting Professionals. Springer, 2020.

[31] P. Karthikeyan, H. M. Pande, and V. Sarveshwaran (eds.), Artificial Intelligence and Blockchain in Digital Forensics. River Publishers, 2023.

[32] G. P. Kumble, Practical Artificial Intelligence and Blockchain: A Guide to Converging Blockchain and AI To Build Smart Applications for New Economies. ʃPackt Publishing, 2020.

[33] B. K. Rai, G. Kumar, and V. Balyan (eds.), AI and Blockchain in Healthcare. Springer 2023.

[34] T. James, Blockchain and Artificial Intelligence: The World Rewired. De Gruyter, 2021.

[35] P. M. S. Choi and S.H. Huang, Fintech with Artificial Intelligence, Big Data, and Blockchain. Springer, 2021.

[36] A. Kumar et al. (eds.), Blockchain, Artificial Intelligence, and the Internet of Things

Possibilities and Opportunities. Springer 2021.

[37] G. Suseendran, R. Anandan, and S. Goundar (eds.), Convergence of Artificial Intelligence and Blockchain Technologies, The Challenges and Opportunities. World Scientific Publishing Company, 2022.

[38] P. M. Tehrani (ed.), Regulatory Aspects of Artificial Intelligence on Blockchain. IGI Global, 2021

[39] P. Fraga-Lamas and T. M. Fernández-Caramés (eds.), Advances in the Convergence of Blockchain and Artificial Intelligence. IntechOpen, 2022.

[40] F. Jaafar and S. Pierre (eds.), Blockchain and Artificial Intelligence-Based Solution to Enhance the Privacy in Digital Identity and IoT. Boca Raton, FL: CRC Press, 2023.

[41] H. Sun, W.Hua, and M. You, Blockchain and Artificial Intelligence Technologies for Smart Energy Systems. Chapman & Hall, 2024.

[42] R. L. Kumar et al. (eds.), Internet of Things, Artificial Intelligence and Blockchain Technology. Springer, 2021.

[43] C. Chakraborty (ed.), Digital Health Transformation with Blockchain and Artificial Intelligence. Boca Raton, FL: CRC Press, 2022.

[44] K. Kilroy, L. Riley, and D. Bhatta, Blockchain Tethered AI. O'Reilly Media, 2023.

[45] S. Gounder, G. Suseendran, and R. Anandan (eds.), The Convergence of Artificial Intelligence and Blockchain Technologies. World Scientific, 2022.

[46] M. Arif et al. (eds.), Artificial Intelligence & Blockchain in Cyber Physical Systems: Technologies & Applications. ⌈Boca Raton, FL: CRC Press, 2023.

[47] W. Spiess-Knafl, Artificial Intelligence and Blockchain for Social Impact : Social Business Models and Impact Finance. Routledge, 2022.

CHAPTER 13

BLOCKCHAIN AROUND THE WORLD

"I think one of the biggest potential promises for governments is credibility enhancement and Blockchain at its core. It fundamentally is a way to minimize the degree of trust that is required, and governments all over the world have historically struggled with maintaining wide trust at all points in time."

— Garrick Hileman

13.1 INTRODUCTION

The world's most important resource is no longer oil, but data. The Internet and smartphones have made data abundant, ubiquitous, and far more valuable. Blockchain is a database technology that records and stores information in blocks of data that are linked, or "chained," together. Data stored on a Blockchain are continually shared, replicated, and synchronized across the nodes in a network. Most Blockchains (and Bitcoin is the biggest) are what we call permission-less systems. With a compound annual growth rate of 56.3%, the Blockchain industry will be worth $163.83 billion by 2029. As Bitcoin attracted considerable amount of attention in recent years, its underlying core mechanism, namely Blockchain technology, has also quickly gained popularity [1].

Due to several factors existing in each country around the world, from economic to regulatory, there are countries that are more ahead of the curve in terms of Blockchain adoption. A practical usage of cryptocurrency around the globe is what differentiates it from every other major currency whether it's euro, dollar, or any other. Numerous country-based financial organizations are putting resources into Blockchain technology to make a more productive organization to deal with economic exchanges.

In recent years, global interest in Blockchain technologies and their possible impact have permeated the public consciousness. Blockchain is essentially software made up of records of digital transactions that are grouped together into "blocks" of information and shared securely across computers on a shared network. Blockchain pilot programs are being implemented around the world in a variety of ways. Numerous nations are presently preferring the utilization of Blockchain technology. Experiments are taking place globally into how BC technology could be used on a national level. The rapid growth of Blockchain has prompted governments around the world to explore ways to regulate it [2].

This chapter summarizes the current Blockchain adoption and regulation in many countries. It begins with providing an overview of Blockchain to make the chapter self-contained. It discusses how Blockchain is implemented around the world. It highlights the benefits and challenges of adopting Blockchain worldwide. The last section concludes with comments.

13.2 OVERVIEW OF BLOCKCHAIN

Blockchain (BC) technology is a permanent record of online transactions. It is a distributed tamper-proof database, shared, and maintained by multiple parties. It is a new enabling technology that is expected to revolutionize many industries, including business. It has the potential for addressing significant business issues. The BC technology allows participants to move data in real-time, without exposing the channels to theft, forgery, and malice.

The term "Blockchain" refers to the way BC stores transaction data – in "blocks" that are linked together to form a "chain." The chain grows as the number of transactions increases. Since every entry is stored as a block on a chain, the care you receive is added to your personal ledger. The first Blockchain was conceived in 2008 by an anonymous person or group known as Satoshi Nakamoto, who published a white paper introducing the concept of a peer-to-peer electronic cash system he called Bitcoin [3].

At its core, Blockchain is a distributed system recording and storing transaction records. In a Blockchain system, there is no central authority. Instead, transaction records are stored and distributed across all network participants. Rather than having a centrally located database that manages records, the database is distributed to the networks and transactions are kept secure via cryptography. BC eliminates the need for a middleman that traditionally may facilitate such transactions. Figure 13.1 shows how Blockchain works [4].

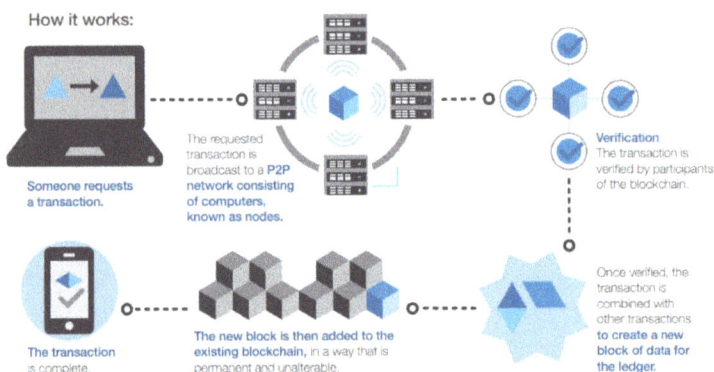

Figure 13.1   How the Blockchain technology works [4].

Fundamentally, Blockchains are distributed digital database that record and maintain a list of transactions taking place in real time. They may also be regarded as decentralized ledgers that sequentially record transactions or interactions among users within a distributed network. Figure 13.2 shows some features of Blockchain [5] Blockchains have the following properties [6]:

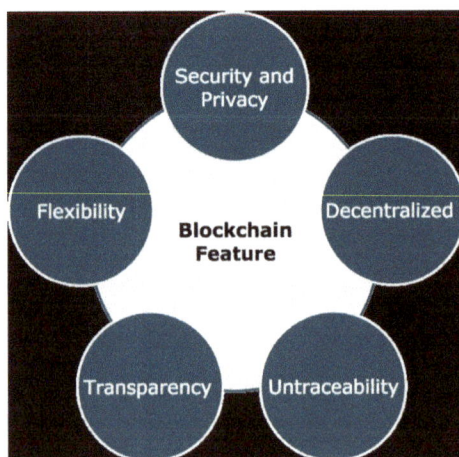

Figure 13.2   Some features of Blockchain [5].

- Firstly, they are autonomous. They run on their own, without any person or company in charge.
- Secondly, they are permanent. They are like global computers

with 100 percent uptime. Because the contents of the database are copied across thousands of computers, if 99 per cent of the computers running it were taken offline, the records would remain accessible and the network could rebuild itself.

- Thirdly, they are secure and tamper-proof. Each record in Blockchain is time stamped and stored cryptographically. The encryption used on Blockchains like Bitcoin and Ethereum is industry standard, open source, and has never been broken.
- Fourthly, they are open, allowing anyone to develop products and services on them.
- Fifthly, as Blockchain is a shared system, costs are also shared between all of its users.

The Blockchain was designed so transactions are immutable, i.e. they cannot be deleted. Thus, Blockchains are secure and meddle-free by design. Data can be distributed, but not copied. When it comes to digital assets and transactions, you can put almost anything on a Blockchain. Different scenarios call for different Blockchains. Blockchain is used in different areas such as depicted in Figure 13.3 [7].

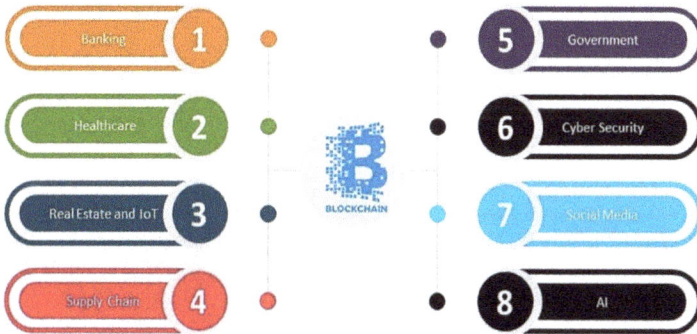

Figure 13.3 Different uses of Blockchain [7].

The BC technology currently has the following features [8,9]:

1. *Peer-to-Peer (P2P) Network*: The first requirement of BC is a network, an infrastructure shared by multiple parties. This can be a LAN at a small scale or the Internet at a large scale. All nodes participating in a BC are connected in a decentralized P2P network. Transactions are broadcast to the

P2P network. Due to some limitations of P2P networks, some vendors have provided cloud-based BCs.

2. *Cascaded Encryption*: A BC uses encryption to protect transaction data. Blocks are encrypted in a cascaded manner, i.e. the encryption result of the previous block is used in encrypting the current block. The BC is secured by public key cryptography, with each peer generating its own public-private key pairs.

3. *Distributed Database*: A BC is digitally distributed across a number of computers. Each party on a BC has access to the entire database and no single party controls the data or the information. Since BC is decentralized, there is no need for central authorizes such as banks.

4. *Transparency with Pseudonymity*: Each node or participant on a Blockchain has a unique 30-plus-character alphanumeric address that identifies it. Users can choose to remain anonymous or provide proof of their identity to others.

5. *Irreversibility of Records*: Once a transaction is entered in the database and the accounts are updated, the records cannot be altered. Records on the database is permanent, chronologically ordered, and available to all others on the network.

There are two types of Blockchains: public and private. Public Blockchains are cryptocurrencies such as Bitcoin, enabling peer-to-peer transactions. Private Blockchains use Blockchain-based platforms such as Ethereum or Blockchain-as-a-service (BaaS) platforms running on private cloud infrastructure. A private BC is an intranet, while a public BC is the Internet. Companies will be disrupted the most by public Blockchains.

13.3 BLOCKCHAIN AROUND THE WORLD

Blockchain is a significant global innovation. Countries around the world are trialing the emerging technology in areas from recording votes in elections to storing the records of citizens. As Blockchain has become a more significant factor in the global investment landscape, countries have taken different approaches to regulate it. Figure 13.4

shows the legal status of bitcoin worldwide [10], while Figure 13.5 displays the venture funding raised by Blockchain [11].

Legal tender

Permissive (legal to use bitcoin)

Restricted (some legal restrictions on usage of bitcoin)

Contentious (interpretation of old laws, but bitcoin is not prohibited directly)

Prohibited (full or partial prohibition)

No data

Figure 13.4   The legal status of bitcoin worldwide [10].

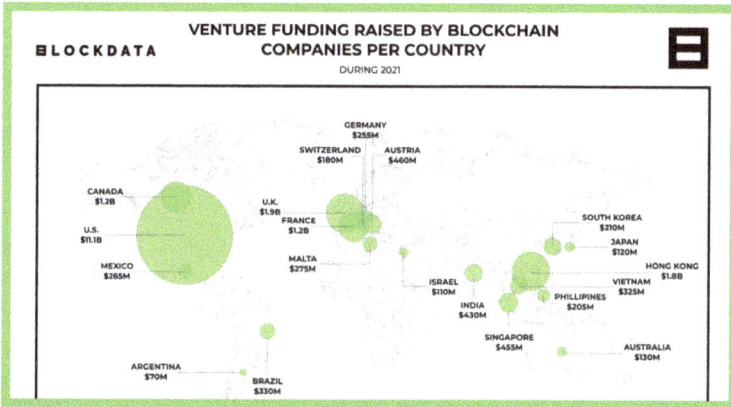

Figure 13. 5   The venture funding raised by Blockchain [11].

Currently, countries like Japan and the United States are topping the rankings for accepting and implementing cryptocurrencies. We will consider how Blockchain is applied in the following nations [12-17].

1. *BLOCKCHAIN IN THE UNITED STATES*

The United States is home to the most extensive crypto ATM network. The country plays a significant role in the adoption of cryptocurrency and Blockchain. Cryptocurrency exchanges are legal in the United States and fall under the regulatory scope of the Bank Secrecy Act (BSA). Nearly 40% of the all Blockchain startup industry is found in the US alone. There has yet to be a cohesive regulation system since laws vary greatly from state to state. The Securities and Exchange Commission (SEC) has already moved toward regulating the sector. We will likely witness US regulators coming down hard on cryptocurrency in the coming years. The present Biden administration seeks to tackle illegal cryptocurrency activity. The government also expects to expand investment in Blockchain development ten times more than before. The US is adopting Blockchain for digital transformation and innovation in a number of departments. The US Navy is developing a supply chain management tool on Blockchain for its logistics. DARPA, the Defense Advanced Research Projects Agency, is investigating the use of Blockchain for secure military communications. The Centers for Disease Control (CDC), has been investigating the use of Blockchain to track public health outbreaks or other medical trends, such as prescription use.

2. BLOCKCHAIN IN EUROPE

In the European Union, the executive agencies recognize the true potential of Blockchain technology and the benefits that it may create for the financial and the governmental sector. Cryptocurrency is widely considered legal, but rules for exchange as well as taxation are different across member states. The European Central Bank is considering the possibility of issuing its own digital currency. It classifies bitcoin as a convertible decentralized virtual currency. In April 2023, Parliament approved measures that allow legislation requiring certain crypto service providers to seek an operating license. This legislation is intended to give regulators the tools they need to track crypto being used for money laundering and terrorism funding. According to the European Central Bank, traditional financial sector regulation is not applicable to bitcoin because it does not involve traditional financial actors. The EU is actively exploring further cryptocurrency regulations. The EU has passed no specific legislation relative to the status of bitcoin as a currency, but has stated that VAT/GST is not applicable to the conversion between traditional (fiat) currency and bitcoin. The European Union has officially launched the European Digital Infrastructure Consortium (EDIC), which will drive Blockchain policy to move Europe's "digital decade" forward.

3. *BLOCKCHAIN IN THE UNITED KINGDOM*

The United Kingdom comes second in the list of countries that possess the most Blockchain-based businesses in the market. While there are no cryptocurrency-specific laws in the UK, the country considers cryptocurrency as property (not legal tender), and crypto exchanges must register with the UK. The UK government views cryptocurrencies as legal tenders, and the assets are treated as foreign currencies. Crypto derivatives trading is banned in the UK as well. There are cryptocurrency-specific reporting. The country is planning to work on its guidelines against identity fraud and delayed monetary services with the appropriate execution of Blockchain technology . Additionally, between 2017 and 2018, more than 500 million euros investments were spent on British-based Blockchain startup companies. The UK government has been exploring Blockchain use cases for many years, seeing its potential in delivering a number of initiatives around safety, trust, transparency, cost and citizen experience. The UK government aims to be at the forefront of Blockchain-related technological innovation for some time to come. The UK Department of Work and Pensions is investigating using Blockchain technology to record and administer benefit payments.

4. *BLOCKCHAIN IN CHINA*

China is one of the top countries using Blockchain technology. It was the first economy to issue its national currency on the Blockchain in early 2021.The country filed massive 225 Blockchain patents in 2017, followed by 91 in the US. People's Bank of China (PBOC) bans crypto exchanges from operating in the country, stating that they facilitate public financing without approval. China placed a ban on Bitcoin mining in May 2021 and on cryptocurrencies in September 2021, forcing many engaging in the activity to close operations. The crackdown issues in China are generally identified with the areas related to Blockchain technology such as fake conduct, token sales, illegal tax laundering, and avoidance of capital controls. Mobile manufacturers in China are racing to launch Blockchain based smartphones. All these instances show that the Chinese government has taken an official interest in Blockchain application. China worked to develop the digital yuan (e-CNY) and in August 2022, it officially rolled out its central bank digital currency (CBDC) pilot test program. China has mastered the art of leading with technologies in recent years. Figure 13.6 shows the percentage of worldwide Blockchain deals of companies based in the US and China [18].

Percent of Worldwide Blockchain Deals of Companies Based in the U.S. and China

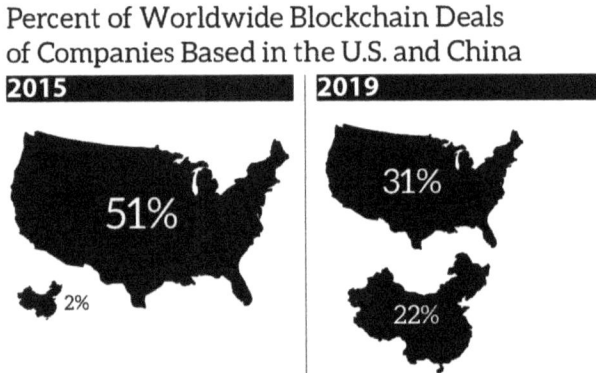

Figure 13.6   The percentage of worldwide Blockchain deals of companies based in the US and China [18].

5. BLOCKCHAIN IN INDIA

India is one of the nations that do not consider cryptocurrencies legal. It remains on the fence regarding crypto regulation, neither legalizing nor penalizing its use. It continues to hesitate to ban crypto outright or to regulate it. The Federal Government in India will encourage Blockchain but is not keen on cryptocurrency trading. Much of the concern in India is about money laundering and other criminal offenses. India continues to be in a dilemma on whether to ban crypto outright or force it into regulation. However, India launched its tokenized rupee pilot program in late 2022.

6. BLOCKCHAINS IN RUSSIA

Right now, it seems like everything related to cryptography in Russia falls under the supervision of national agencies such as the Federal Security Service and others. Although the certification is not legally required, it is somewhat suggested that transactions between Blockchain participants will have no legal significance if they are not certified. Russia has not always been clear on their stance towards crypto, but they have openly embraced blockchain technology in the country. The popular Russian voting service Active Citizen utilizes a government blockchain protocol that allows residents to vote. In 2020, Russian President Vladimir Putin signed a law that regulates digital financial asset transactions. Under the law, digital currencies are recognized as a payment means and investment.

7. BLOCKCHAIN IN SWITZERLAND

Switzerland tops the list of countries using Blockchain technology due to its establishment of "Crypto Valley." Zug, a town located in the very heart of Switzerland is now termed as the "Crypto Valley" due to large amounts of investments specifically in the Blockchain sphere. The tiny town has more than 450 blockchain-integrated businesses and associations. The Swiss Federal Council has adopted a range of supportive laws and now Blockchain is widely used in a range of financial transactions. The country also enjoys the status of a tax-free haven for cryptocurrency investors.

8. BLOCKCHAIN IN ESTONIA

This is one of the first nation to use Blockchain on national level. The country in itself is a tech organization. It has one of the most progressive views of using blockchain. The country has transformed itself into a virtual society that deploys new technologies to improve the economy and way of living for the citizens. The country is planning to deploy BC technology in governance, banking services, healthcare, and other Bitcoin-based ATMs. The Ministry of Finance in Estonia does not place any obstructions on the use of cryptocurrencies in payment transactions. Estonia is also the first to introduce e-voting, based on blockchain technology. Anyone who wants to live in Estonia will receive an e-citizenship from the government.

9. BLOCKCHAIN IN UNITED ARAB EMIRATES

Dubai is the capital of UAE and is a forerunner in Blockchain adoption and one of the hottest real estate markets for the investors. The country adopted Blockchain technology as early as 2016 and is planning to be the first Blockchain-powered government by 2020.
Dubai has been a global pioneer in Blockchain adoption, and has already experienced the multi-faceted benefits that this technology can bring.

The UAE Government launched the Emirates Blockchain Strategy 2021 to capitalize on Blockchain technology, with Smart Dubai spearheading the Emirate's Blockchain transformation. Digital Dubai was established in June 2021 to develop and oversee the implementation of policies and strategies that govern all matters related to Dubai's information technology, data, digital transformation, and cyber-security. It seeks to make Dubai the happiest city on earth.

10. *BLOCKCHAIN IN CANADA*

While crypto is not considered legal tender in Canada, the country has been more proactive than others about crypto regulation. Canada became the first country to approve trading Bitcoin on the Toronto Stock Exchange. In Canada, a strong focus can be seen in utilizing Blockchain for identity and credentials management. The Canadian government has already piloted Blockchain technology for its own digital credential management system. Canada is testing the Known Traveller Digital Identity System in collaboration with the World Economic Forum and the Netherlands. The program uses biometrics, cryptography and Blockchain to let travelers control and share information. In 2021, the Canadian Securities Administrators (CSA) published guidance for crypto issuers that own or hold crypto assets.

11. *BLOCKCHAINS IN AUSTRALIA*

Tourism is one of Queensland's most important industries and among the major source of revenue generation. Cryptocurrencies as well as Bitcoin are classified as property, subjecting them to capital gains tax. Exchanges are allowed to operate in the country, provided that they register with the Australian Transaction Reports and Analysis Centre. This opens many doors for investors, considering the fact that even Australian government provides specific funds for development and improvement of Blockchain standards in this country. The government has just started to get involved in this sector and has come up with a national Blockchain roadmap as part of its big plan. This five-year plan is expected to bring certain regulatory mechanisms. From this, we can see that Australia is one of the most progressive nations in this field. In 2021, Australia announced plans to create a licensing framework around cryptocurrency and potentially launch a central bank digital currency. Food fraud, which costs Australia billions each year, could also be addressed with a Blockchain-based supply chain. Centralized health records could be made easily accessible, safe and private via a Blockchain solution. In the state of South Australia, an election to appoint a government advisory council was conducted using Blockchain technology, illustrating how Blockchain might one day be used to broaden Blockchain voting mechanisms and enhance citizen input.

12. *JAPAN*

Japan is among the early adopters of Blockchain technology. Japan has been a crypto-center from the beginning. It takes a progressive approach to crypto regulations. The innovator of Bitcoin is said to be Japanese due to his name "Satoshi Nakamoto." The Japanese government empowers the implementation of Blockchain technology, attempting to lead in worldwide innovative developments. Japan takes a progressive approach to crypto regulations. Right now, Japan is the only nation to have legitimate lawful cryptocurrency guidelines. In April 2017, The Tokyo government passed a law perceiving Bitcoin as a legal currency. Further, as indicated by the Japanese Financial Services Agency (FSA), over 3.5 million people exchange cryptocurrencies and accept the digital asset as an actual asset.

13. *BLOCKCHAIN IN SINGAPORE*

This island state is an aspiring Smart Nation with its robust methodology that tries to change this previously known fishing town into a hub of the research facility. It classifies cryptocurrency as property but not legal tender. The Singapore government is perceived as more business-accommodating and transparent in comparison to other countries. Singapore is a leading country in Blockchain adoption, with the government investing heavily in Blockchain research and development, causing many Blockchain businesses to choose to incorporate there. It supports a few new Blockchain-based businesses. Their financial regulatory body, the Monetary Authority of Singapore (MAS), encourage Blockchain and cryptocurrency adoption. Its primary role is license and regulate exchanges and monitor and mitigate the crypto industry risks without hindering technological innovation. MAS has launched a S$12 million Singapore Blockchain Innovation Programme

(SBIP) in the continued effort to build the country's Blockchain ecosystem. Singapore gets its reputation as a cryptocurrency safe haven because long-term capital gains are not taxed. It has already developed a number of Blockchain concepts, such as for inter-bank and cross-currency payments with Europe and Canada. When it comes to Blockchain advancement and applications, it has a neck-to-neck rivalry with China and Japan.

14. *BLOCKCHAIN IN NIGERIA*

The country is the most populous African nation of almost 200 million people. It is among the top countries with a high global crypto adoption index. The Nigerian population faces an uncertain domestic situation, with high inflation and depreciation of the country's fiat currency. African nations are no strangers to the use of digital solutions for money transfers. In early 2017, the Central Bank of Nigeria warned financial institutions not to use, hold or trade virtual currencies pending "substantive regulation or decision by the (Central Bank of Nigeria) as they are not legal tender in Nigeria." Nigeria has a huge commercial market for crypto. More and more commerce is done on the rails of cryptocurrency, including international trade with counter parties in China. The adoption of Blockchain and cryptocurrency has experienced a clear uptrend in Nigeria, especially after the 2022 crypto market crisis when the country emerged as one of the most crypto-curious nations. In October 2021, the Central Bank of Nigeria launched the eNaira, a Blockchain-based central bank digital currency (CBDC) pegged to the country's national currency, the naira.

13.4 BENEFITS

In countries with historically weak currencies, including several Latin American and African countries, Bitcoin has become popular with populist leaders. Law enforcement agencies and regulatory bodies around the world are working hard, trying to develop strict guidelines for the safe and reliable deployment of these digital assets. Blockchain can enable businesses to solve issues with real-time data access, partners' privacy, and traceability. Other benefits include the following [19,20]:

- *Security*: Regardless of the business type, Blockchains are necessary for maintaining a high level of security. Blockchain technologies can prevent identity theft, as well as be beneficial for the protection of data integrity. Traditionally, documents are signed with private keys, but the process is complicated and requires that the keys be proven to have not been manipulated. Blockchain technology serves as an alternative in this case.

- *Decentralization*: This constitutes one of the key ideas behind blockchain technology. With a decentralized distribution of information, data security gets a boost. When information in the system is distributed through internal servers, it prevents unauthorized access to the data. Consequently, the chances of a successful cyber attack leading to information theft are drastically reduced.

- *Remittances*: Emigrants working abroad send billions of dollar to their families in their home countries. But the process of sending money can be extremely expensive. Some organizations are using Blockchain technology to reduce the cost of remittances transferred across borders by migrant workers. Currently it is estimated that at least $32 billion in remittances is failing to reach recipients, due to high transaction fees associated with sending and receiving money internationally.

- *Digital Identity*: According to the United Nations, one in every five people globally lacks a legal identity. The World Identity Network and Humanized Internet project can store identifiers such as birth certificates and university degrees on a Blockchain. Users can keep their information private and secure, but also give permission for anyone to access it anywhere in the world.

- *Land Ownership Rights*: A groundbreaking use of the Blockchain is for securing land ownership rights. Proof of land ownership is a challenge in many parts of the developing world. One organization, Bitland, is piloting a project in Ghana to provide services that allow individuals and groups to survey land and record title deeds on a Blockchain.

- *Governance and Democracy*: Government and civil

society can leverage Blockchain technology to strengthen democratic processes and citizens participation. Blockchain systems such as Ballotchain can manage online elections. The system ensures that voters cannot vote twice or commit electoral fraud, thus ensuring the integrity of election processes.

• *Helping the Poor*: The real benefits of Blockchain technology could come to the world's poorest people. Blockchain is an inexpensive and transparent way to record transactions. The technology enables people to securely exchange money without fear of fraud or theft. This will help them get easier access to banks for loans or to protect their savings. Blockchain technology can also improve humanitarian assistance. Fraud, corruption, discrimination, and mismanagement hinder some money intended to reduce poverty and improve education and health care from actually helping people.

• *Insurance*: Most people in the developing world lack health and life insurance, primarily because it is expensive. They have neither the ability to afford any insurance nor any company offering them services. Because blockchain systems are online and involve verification of transactions, they can deter fraud, dramatically cutting costs for insurers.

• *Quality Assurance*: Maintaining a high level of quality is a key priority for any business. It is important to maintain the integrity of the supply chain and provide the customer with services that meet their standards. Blockchains can be used in industries for detecting irregularities. For example, BC is being incorporated into the food sector, where businesses ensure the safety of food by tracking batch information.

13.5 CHALLENGES

In spite of the potential benefits of Blockchain, it is still a young technology and presents a number of challenges. While cryptocurrency has existed since 2009, governments and regulators are still working out ways to govern its uses. The studies have shown that most developing and well-established nations are dominating the space of cryptocurrency and Blockchain technology. Financial inclusion is a challenge to developing nations. Governments around the world are facing new policy and regulatory challenges, not only in ensuring compliance but also in managing issues arising from digital disruption. Other challenges include the following [18]:

- *Safety*: The main concern about Blockchain remains its safety. Hackers have stolen millions of dollars in Bitcoin, using only phone numbers. This type of security weakness can be used against anyone who is using their phone for services as Google, PayPal, Dropbox, Facebook, Twitter, and iCloud.

- *Complexity*: BC relies on complicated authentication algorithms and cryptography.

- *Immutability*: Another challenge comes from the immutable nature of Blockchain. If a disagreement arises between two parties involved in a transaction, there is no way to go backward. Modifications require that the network create an additional record (or block) to confirm a change. If individuals cannot agree on a change, they are stuck with the original agreement forever.

- *Public Perception*: Most people remain skeptical about the libertarian utopia of Blockchain and the radical philosophy it presents. Leaders across industries have seemed unsure what to do with Blockchain technology.

- *Uneven Adoption*: The adoption of Blockchain has not been equal among the different nations. Some nations have made giant strides in the Blockchain space, while others are still playing catch up. Some authorities are cautious about the emerging technology, and are reluctant to adopt supportive laws, while others recognize the advantages of the Blockchain, and continue to issue favorable legislations.

- *Regulation*: The lack of a robust global regulatory framework for digital assets is harmful for the sector. Blockchain has expanded to an extent that can no longer be ignored by regulatory authorities. Every nation develops on its own regulations on Blockchain. Mass Blockchain adoption will require a combination of global innovation and cooperation. In the US, the federal government is reluctant to accelerate the adoption of laws that support Blockchain, and the policies continue to be issued only by separate states. Figure 13.7 shows cryptocurrency regulation in the US [19].

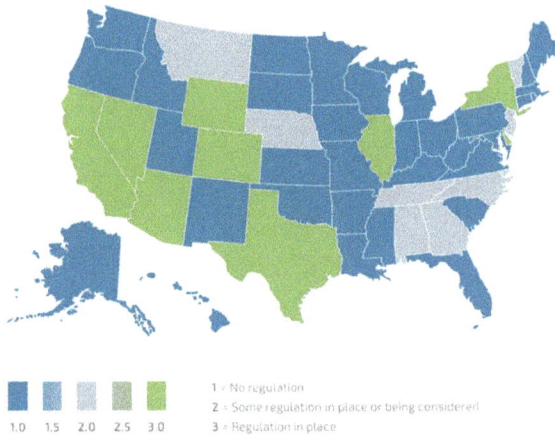

Figure 13.7 Cryptocurrency regulation in the US [19].

13.6 CONCLUSION

Blockchain is the subject of extensive international interest and attention over the past few years. The global rate of adoption of cryptocurrencies has skyrocketed in recent times. For every country, there is a government trialing a Blockchain pilot. Blockchain universities around the world are advancing the careers of tech aspirants.

Learning Blockchain is easy now due to reputed online courses and training sessions. Blockchain Council is one such organization that offers online training and certification programs to aspiring trainees. Global Blockchain Business Council (GBBC) is the largest leading industry association for the Blockchain technology and digital assets community. Launched in Davos in 2017, GBBC is a Swiss-based non-profit, with more than 500 institutional members. More information about Blockchain around the world can be found in the books in [21-24].

REFERENCES

[1] J. Shangrong et al., "Policy assessments for the carbon emission flows and sustainability of Bitcoin Blockchain operation in China," Nature Communications, vol. 12, no.1, April 2021.

[2] M. N. O. Sadiku, U. C. Chukwu and J. O. Sadiku, "Blockchain around the world," International Journal of Human Computing Studies, vol. 5, no. 9, September 2023, pp.1-10.

[3] M. N. O. Sadiku, Y. Wang, S. Cui, and S. M. Musa, "A primer on Blockchain," International Journal of Advances in Scientific Research and Engineering, vol. 4, no. 2, February 2018, pp. 40-44.

[4] "Building Block(chain)s for a better planet,"

https://www3.weforum.org/docs/WEF_Building-Blockchains.pdf

[5] https://www.researchgate.net/figure/characteristics-of-Blockchain-the-signed-data-As-long-as-the-user-prevents-the-private_fig2_339084601

[6] S. Depolo, "Why you should care about Blockchains: The non-financial uses of Blockchain technology," March 2016,

https://www.nesta.org.uk/blog/why-you-should-care-about-Blockchains-non-financial-uses-Blockchain-technology

[7] O. Bheda, "What is Blockchain?" https://builtin.com/ Blockchain

[8] M. Iansiti and K. R. Lakhani, "The truth about Blockchain," Harvard Business Review, Jan./Feb. 2017.

https://hbr.org/2017/01/the-truth-about-Blockchain

[9] W. T. Tsai et al., "A system view of financial Blockchains," Proceedings of IEEE Symposium on Service-Oriented System Engineering, 2016, pp. 450-457.

[10] "Legality of cryptocurrency by country or territory," Wikipedia, the free encyclopedia

https://en.wikipedia.org/wiki/Legality_of_cryptocurrency_by_country_or_territory

[11] "Top 20 countries ranked by Blockchain investment in 2021,"

https://www.linkedin.com/pulse/top-20-countries-ranked-Blockchain-investment-2021-colangelo

[12] "Cryptocurrency regulations by country,"

https://www.thomsonreuters.com/en-us/posts/wp-content/uploads/sites/20/2022/04/Cryptos-Report-Compendium-2022.pdf

[13] D. Perry, "Regulation of cryptocurrency around the World – 2023 Guide," September 2020,

https://revenuesandprofits.com/cryptocurrency-around-the-world/

[14] K. George, "Cryptocurrency regulations around the world," July 2023,

https://www.investopedia.com/cryptocurrency-regulations-around-the-world-5202122

[15] T. K. Sharma, "Top 10 countries leading Blockchain technology in the world," December 2022,

https://www.Blockchain-council.org/Blockchain/top-10-countries-leading-Blockchain-technology-in-the-world/

[16] E. Lacapra, "5 countries leading Blockchain adoption," March 2023,

https://cointelegraph.com/news/5-countries-leading-the-Blockchain-adoption

[17] H. Cooper, T. Hill, and S. Kangalingam, "Around the world in government Blockchain," August 2022,

https://www.pwc.com.au/digitalpulse/government-Blockchain-use-cases.html

[18] T. Frohwitter, "While the world battles the novel coronavirus, China is transforming into the global blockchain leader," May 2020,

https://fordhamobserver.com/46498/opinions/while-the-world-battles-the-novel-coronavirus-china-is-transforming-into-the-global-blockchain-leader/

[19] "Digital currencies and Blockchain in the social sector," January 2018,

https://ssir.org/articles/entry/digital_currencies_and_Blockchain_in_the_social_sector1#

[20] N. Kshetri, "Can blockchain technology help poor people around the world?" April 2017,

https://theconversation.com/can-blockchain-technology-help-poor-people-around-the-world-76059

[21] D. Tapscott and A. Tapscott, Blockchain Revolution: How the Technology Behind Bitcoin and Other Cryptocurrencies is Changing the World. Portfolio, 2018.

[22] P. Domjan et al., Chain Reaction: How Blockchain Will Transform the Developing World. Springer,2021

[23] IntroBooks Team, How Will Blockchain Change The World. IntroBooks Team, 2020.

[24] Progressive Management, 2018 Complete Guide to Regulation of Cryptocurrency Around the World: Survey of 130 Countries and Organizations - Bitcoin, Virtual Currencies, Digital Money, Blockchain Technologies Laws and Policies. Apple Books, 2018.

CHAPTER 14

FUTURE OF BLOCKCHAIN

"It looks like Blockchain is here to stay, I think it's going to be a powerful technology for modern society."
— Reid Hoffman

14.1 INTRODUCTION

The proliferation of the Internet has profoundly altered traditional businesses, causing some to decline while others are growing. The Internet has generated significant changes in the landscape of various jobs and businesses and has given rise to many new and innovative business models. The digital economy has been shaped by the development of the Internet and cyberspace, giving rise to various technologies such as Blockchain technology. Undoubtedly, Blockchain is one of the most important technologies that has emerged in the last decade [1].

Blockchain is revolutionizing the digital world by bringing a new perspective to security, efficiency, and stability of systems and data. It is network of computers that is decentralized. Blockchain is a decentralized system that keeps track of distributed data and provides encrypted transaction tracking. It has attracted attention with its unique characteristics, such as irrevocability and security. It will be a part of our everyday life.

Every Blockchain features a consensus to assist the network make decisions and consensus algorithms are at the core of its architecture. Blockchain aims to achieve a decentralized ledger. Because of its decentralization, security, and irrevocable characteristics, Blockchain is widely used in cryptocurrencies and other fields [2].

Although it is difficult to predict the future technology landscape and exactly how Blockchain fits in, but this chapter has identified a few areas of potential impact. The future of Blockchain will be determined by how effectively Blockchain integrates with advanced technologies, such as artificial intelligence (AI), Internet of things (IoT), and 5G networking.

This chapter attempts to predict the future of Blockchain technology. It commences by providing an overview of Blockchain to make the chapter self-contained. The chapter explores the future of Blockchain by first considering novel technologies with the potential to facilitate future development of Blockchain, and then delving into its implications in different applications. Such technologies include Internet of things and artificial intelligence. It covers future applications of Blockchain. It highlights the benefits and challenges Blockchain will face in the future. The last section concludes with comments.

14.2 OVERVIEW OF BLOCKCHAIN

Blockchain (BC) technology is a permanent record of online transactions. It is a distributed tamper-proof database, shared, and maintained by multiple parties. It is a new enabling technology that is expected to revolutionize many industries, including business. It has the potential for addressing significant business issues. The BC technology allows participants to move data in real-time, without exposing the channels to theft, forgery, and malice.

According to its name, the Blockchain is a combination of the two terms "block" and "chain." So a "Blockchain" refers to the way BC stores transaction data in "blocks" that are linked together to form a "chain." "Block" refers to blocks of data, involving transactions, while "chain" refers to chain-like linking of blocks to previous ones until the genesis block is reached. The chain grows as the number of transactions increases. Since every entry is stored as a block on a chain, the care you receive is added to your personal ledger. The first Blockchain was conceived in 2008 by an anonymous person or group known as Satoshi Nakamoto, who published a white paper introducing the concept of a peer-to-peer electronic cash system he called Bitcoin [3]. A Blockchain layered architecture is shown in Figure 14.1 [4].

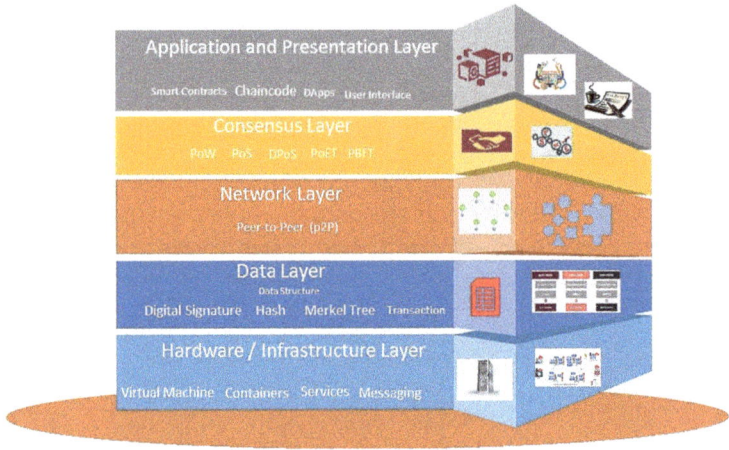

Figure 14.1   The Blockchain layered architecture [4].

At its core, Blockchain is a distributed system recording and storing transaction records. In a Blockchain system, there is no central authority.   Instead, transaction records are stored and distributed across all network participants. Rather than having a centrally located database that manages records, the database is distributed to the networks and transactions are kept secure via cryptography. BC eliminates the need for a middleman that traditionally may facilitate such transactions. Figure 14.2 shows how Blockchain works [5].

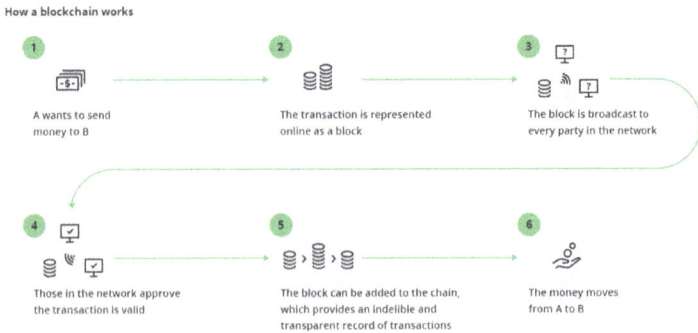

Figure 14.2   How the Blockchain technology works [5].

Fundamentally, Blockchains are distributed digital database that record and maintain a list of transactions taking place in real time. They may also be regarded as decentralized ledgers that sequentially record transactions or interactions among users within a distributed network. They have the following properties [6]:

- Firstly, they are autonomous. They run on their own, without any person or company in charge.

- Secondly, they are permanent. They are like global computers with 100 percent uptime. Because the contents of the database are copied across thousands of computers, if 99 per cent of the computers running it were taken offline, the records would remain accessible and the network could rebuild itself.

- Thirdly, they are secure and tamper-proof. Each record in Blockchain is time stamped and stored cryptographically. The encryption used on Blockchains like Bitcoin and Ethereum is industry standard, open source, and has never been broken.

- Fourthly, they are open, allowing anyone to develop products and services on them.

- Fifthly, as Blockchain is a shared system, costs are also shared between all of its users.

The Blockchain was designed so transactions are immutable, i.e. they cannot be deleted. Thus, Blockchains are secure and meddle-free by design. Data can be distributed, but not copied. When it comes to digital assets and transactions, you can put almost anything on a Blockchain. Different scenarios call for different Blockchains. Blockchain is used in different areas such as depicted in Figure 14.3 [5]. It is explored by several organizations as depicted in Figure 14.4 [7].

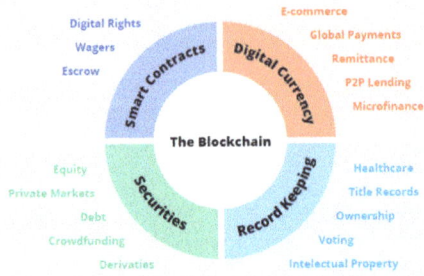

Figure 14.3   Potential uses of Blockchain [5].

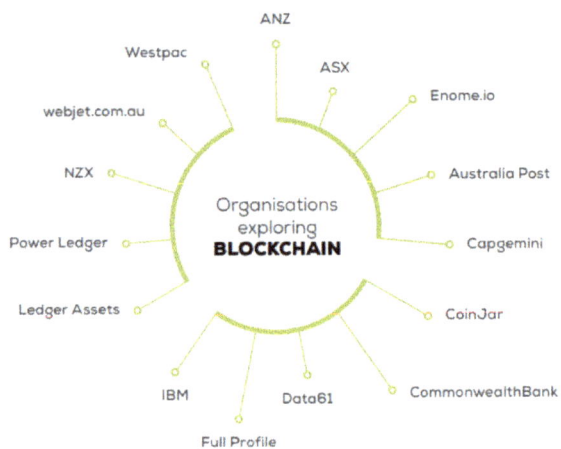

Figure 14.4   Organizations exploring Blockchain [7].

The BC technology currently has the following features [8,9]:

1. *Peer-to-Peer (P2P) Network*: The first requirement of BC is a network, an infrastructure shared by multiple parties. This can be a LAN at a small scale or the Internet at a large scale. All nodes participating in a BC are connected in a decentralized P2P network. Transactions are broadcast to the P2P network. Due to some limitations of P2P networks, some vendors have provided cloud-based BCs.

2. *Cascaded Encryption*: A BC uses encryption to protect transaction data. Blocks are encrypted in a cascaded manner, i.e. the encryption result of the previous block is used in encrypting the current block. The BC is secured by public key cryptography, with each peer generating

its own public-private key pairs.

3. *Distributed Database*: A BC is digitally distributed across a number of computers. Each party on a BC has access to the entire database and no single party controls the data or the information. Since BC is decentralized, there is no need for central authorizes such as banks.

4. *Transparency with Pseudonymity*: Each node or participant on a Blockchain has a unique 30-plus-character alphanumeric address that identifies it. Users can choose to remain anonymous or provide proof of their identity to others.

5. *Irreversibility of Records*: Once a transaction is entered in the database and the accounts are updated, the records cannot be altered. Records on the database is permanent, chronologically ordered, and available to all others on the network.

There are two types of Blockchains: public and private. Public Blockchains are cryptocurrencies such as Bitcoin, enabling peer-to-peer transactions. Private Blockchains use Blockchain-based platforms such as Ethereum or Blockchain-as-a-service (BaaS) platforms running on private cloud infrastructure. A private BC is an intranet, while a public BC is the Internet. Companies will be disrupted the most by public Blockchains.

14.3 BLOCKCHAIN FUTURE TRENDS

There is no doubt that the Blockchains that support the currencies will help change our society in countless ways. We will discuss the current problems and opportunities of Blockchain in combination with the development of other technologies. As the adoption of Blockchain technologies becomes more widespread, industry leaders envision Blockchain integrating with advanced technologies, such as artificial intelligence (AI), Internet of things (IoT), and 5G networking. The following sections present novel technologies with the potential to facilitate future development of Blockchain.

14.4 FUTURE OF INTERNET OF THINGS

Internet of things (IoT) consists of smart electronic devices, local area networks, the Internet, cloud servers, and the user application. It has emerged as a leading technology worldwide. The technology has become a mainstream technology and is showing no signs of slowing down in terms of innovation. No business can grow without the implementation of IoT. The Internet of things (IoT) interconnects embedded devices using sensors through private or public networks, reduces human intervention as much as possible, and realizes seamless communications among people, products, and processes.

IoT devices may be simultaneously managed by multiple managers; thus, the centralized network architecture faces the risk of privacy disclosure. Blockchain provides a new idea and solution to IoT challenges considering scalability, cooperation ability, trust relationship, security, and privacy protection. All data transmitted through Blockchain are encrypted, and user data and privacy are more secure. As long as the data is written into the Blockchain, tampering with it is difficult [10]. Blockchain technology has the potential to provide a secure and scalable framework for communication between IoT devices. With a higher resistance to cyber-attacks than existing IoT security solutions, Blockchain will also allow smart devices to quickly and cost-effectively make automated transactions. In a world ruled by gadgets and the Internet of things (IoT), security and integrity are major concerns for users. The future of the Internet of things is looking bright with amazing opportunities and with new technologies and access to information that we may not previously have thought possible. The potential of the IoT devices is enormous and remains untapped. IoT is about to expand to limits unsurpassed by anything yet [11]. In the future, IoT can potentially undergird the infrastructure of smart cities in order to make communications far more streamlined and efficient than they are today.

14.5 FUTURE OF ARTIFICIAL INTELLIGENCE

Blockchain and artificial intelligence (AI) are two of the most transformative and disruptive technologies of our times. All major economies in the world are racing to take the lead in the development and deployment of AI and Blockchain technologies. Blockchain is considered a shared and permanent ledger that will be used for the encryption of data in the future. AI refers to machines capable thinking like humans, mimicking their actions, learning problem solving, rationalizing, and taking actions that have the best chance of achieving a specific outcome. The proliferation of AI and robots has significantly impacted contemporary businesses and human lives. The combination of AI and emerging Blockchain-based technologies is expected to generate significant advancements. It is often credited with ushering in the fourth industrial revolution, with their implementation revolutionizing industries such as healthcare, finance, supply chain, and logistics. The implementation of AI-enabled Blockchain technology raises several challenges, such as the cost of implementation and the availability of skilled personnel to develop and manage such systems. The integration of AI and Blockchain technology into existing business processes and systems may require significant changes in an organization's culture and procedures. In the future, the integration of could transform a host of industries. The potential of AI and Blockchain technology is limitless [12,13].

14.6 FUTURE BLOCKCHAIN APPLICATIONS

Blockchain technology has proved its place as the technology of the future. Many experts predict that Blockchain technology has the potential to revolutionize various industries such as healthcare, supply chain, logistics, voting, banking, real estate, and insurance. These applications could serve as the foundation for the next generation of businesses. Future applications of Blockchain include [14,15]:

- *Cryptocurrencies*: These are digital currencies that provide an alternative payment method. Cryptocurrency and Blockchain are closely related because cryptocurrencies rely on Blockchain technology to operate. In fact, Blockchain has been the key driver behind the rise of cryptocurrency because of its basic features of decentralized approach, enhanced security, and immutable ledgers. Cryptocurrency, such as Bitcoin and Ethereum, is a popular financial asset that has gained worldwide adoption in recent times. With the help of Blockchain technology, transactions can be conducted securely and transparently, without the need for intermediaries such as banks. In the coming years, there will be a significant increase in the adoption of cryptocurrencies. This would lead to more widespread use of Blockchain technology as businesses begin to accept cryptocurrency payments. Cryptocurrency could be a threat to the traditional banking industry. It offers what banks offer today but with higher security and less manual work. Cryptocurrency is the future of money. Figure 14.5 shows ten popular types of cryptocurrency [16].

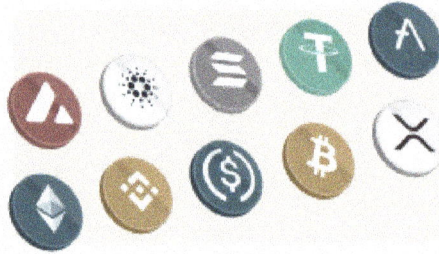

Figure 14.5 Ten popular types of Cryptocurrency [16].

The four emerging coins that one should be investing in are Bitcoin, Etherium, NEO, and EOS. China is practically prepared to give their Crypto yuan.

- *Non-Fungible Tokens*: Although cryptocurrencies are one of the most widely-known use cases of Blockchain, another emerging use case of Blockchain technology is the non-fungible token (NFT). Instead of selling or buying physical assets, NFTs enable you to buy and sell digital assets that represent real-world items. All NFTs are unique and cannot be replaced or swapped; they can only be purchased, sold, traded, or given away by the original owner/creator of that asset. NFTs are always available to be sold and bought. The system for NFTs never goes down. Figure 14.6 shows examples of fungible and non-fungible tokens [17].

Examples of Fungible & Non-Fungible Tokens

Fungible	Non-Fungible
Dollar	Cryptokitties
Bitcoin	Art
Ethereum	House/Property

Figure 14.6 Examples of fungible and non-fungible tokens [17].

- *Finance*: Finance is naturally the industry where Blockchains will have the most impact. Blockchain technology stands to revolutionize the way money can be handled. In the financial services sector, Blockchain technology has already been implemented in many different and innovative ways. Blockchain is a distributed peer-to-peer ledger system that eliminates the need for a centralized management entity such as a bank. Brokers are offering crypto as tradable assets. Banks and financial institutions from all over the world have realized how useful Blockchains are for a range of reasons. Today, Blockchains are used for sending large amounts of money across the world instantaneously and with minimum fees.

- *Healthcare*: The healthcare industry can use Blockchain technology to store and share patient information between different healthcare providers. Imagine a world where all your medical records were readily available wherever you are, and where medicines and prescriptions were handled on a completely safe platform. Blockchain has the potential to make all this happen. It can play a key role within the healthcare sector by increasing the privacy, security, safety, convenience, and interoperability of the healthcare data. It can also limit the number of mistakes made by health care professionals. Blockchain can be used to develop applications to manage patient data, control drug supply, automate medical examination, treatment transactions, and more.

- *Voting*: Transparency and autonomy are two Blockchain features that motivated developers to apply this technology to electronic voting systems. Transparency enables the public to monitor the voting process, and autonomy removes the influence of authorities [1]. Voting with Blockchain technology could be easier, faster, and more secure than how we vote today. Blockchain technology can end voter fraud. It offers the power to vote digitally, but it is transparent enough that any regulators would be ready to see if something were changed on the network. It combines the convenience of digital voting with the immutability to make your vote truly count. People can vote online easily without revealing their identities.

- *Supply Chain*: Blockchain technology can trace all the steps of a supply chain. Through the use of Blockchain technology, manufacturers can identify the sources of goods, deliveries, and production activities all through a supply chain management process. Blockchain technology has already been used in multiple industries as a means of keeping tabs on supply chains and ensuring their efficiency. Blockchain technology can trace all the steps of a supply chain.

- *Cybersecurity*: Blockchain technology can secure our data against unauthorized access and tampering. It is ideal for environments where high security is involved. You can easily identify malicious data attacks with Blockchain due to peer-to-peer connections, where data cannot be altered or tampered. And, by eliminating a central authority, Blockchain provides a secure and transparent way of recording transactions without disclosing private information to anyone.

- *Government*: Blockchains are global networks that can have millions of users, each adding data which is secured through cryptography. It is envisaged that governments will develop their own digital money and engage in free market trading. Other applications for government include digital asset registries, wherein the fast and secure registry of an asset such as a car, home or other property is needed; notary services, where a Blockchain record can better verify the seal's authenticity; and taxes, in which Blockchain technology can make it easier to enable quicker tax payments. The future application of Blockchain technology possibly is a nation's digital money. While the exploitation of Blockchain technology can yield tremendous benefits, not all nations are on the same page with regard to adopting it. Figure 14.7 shows a worldwide adoption of Blockchain [5].

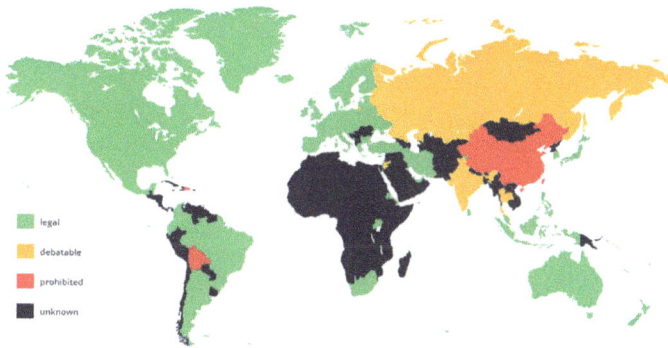

Figure 14.7   Worldwide adoption of Blockchain [5].

- *Digital Identity*: Today, we use passwords and authentication questions to prove who we are online. Blockchain could replace this system with a digital identity that is safe, secure, and easy to manage. Your digital identity is based on the uniquely random set of numbers assigned to each user on a Blockchain network. This implies that your identity cannot be hacked or changed without access to your private key, which makes it more reliable than our current solution. Digital identities could be stored and managed through Blockchain technology and this applies to health records as well. Blockchain has the ability to offer a cross-border identity standard for all the citizens of the world. Blockchain for digital identity will also be a prominent sector in the future.

Some of these future applications of Blockchain are depicted in Figure 14.8 [18].

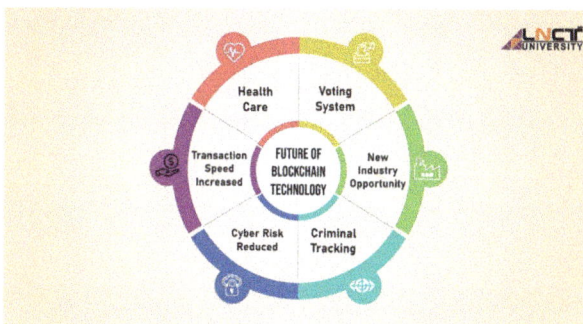

Figure 14.8   Some future applications of Blockchain [18].

14.7 BENEFITS

Without a doubt, the intrinsic benefits of Blockchain technology are difficult to ignore. Blockchain may potentially be used to securely and efficiently transfer user data across platforms and systems. Nobody owns or controls your data but you. Blockchain has the potential to promote transparency, efficiency, and responsibility. This ensures that Blockchain technology is used in a way that benefits both people and the planet. The technology is poised to play a key role in shaping a sustainable future. Blockchain talent became one of the most sought-after skill sets globally. Other benefits include the following [19]:

- *Decentralization*: This refers to the distribution of information, control, and ownership from a single organization such as a government or administration to multiple nodes scattered across a network. Decentralization is one of the most important aspects of Blockchain, particularly in financial transactions as it eliminates the need for intermediaries like banks and financial institutions.

- *Transparency*: This is another crucial feature of Blockchain technology. All transactions are recorded on a publicly available ledger, allowing for complete transparency of data. This enables users to view and verify the transactions taking place on the network. Transparency makes it easy to track the movement of assets and provides a reliable and trustworthy source of information. This helps to prevent fraudulent activities and reduce the potential for corruption within the system. Blockchain is enabling higher transparency in the supply chain for various goods.

- *Security*: Blockchain offers a very high level of security through the independent verification processes that take place throughout the member computers on a Blockchain network. Due to the secured and decentralized nature of Blockchain technology, it is very difficult, if not impossible, for hackers or nefarious parties to tamper with transactions. The principles of cryptography, decentralization, and consensus mechanisms contribute to making data secure when stored on Blockchain technology. All Blockchain transactions are encrypted so it gives the network high integrity. The Blockchain system uses asymmetric cryptography to secure transactions between users through the help of public and private keys. The use of public and private keys in Blockchain technology provides a secure way to transfer assets and information on the network. Once a transaction is added to the distributed ledger, it becomes extremely difficult to alter or manipulate data on the Blockchain, ensuring the integrity and security of the system.

• *Smart Contracts*: These are self-executing digital agreements that guarantee the fulfillment of a contract. Lawyers are utilizing Blockchain to create smart contracts. The primary thought of smart contracts is its programmed execution when conditions are met. An example is conveying products after payment is received. These contracts are stored on a Blockchain network, which automates the execution of the contract's predefined conditions. They are visible to everyone, including the users, providing a layer of transparency and ensuring the trustworthiness of the data. Blockchain technology utilizes smart contracts to simplify business and trade between both anonymous and identified parties, with or without the need for a middleman. Smart contracts have the ability to automate contract execution and reduce costs. They have a wide range of applications, including supply chain management, insurance, real estate, and more. Figure 14.9 shows smart contract [20].

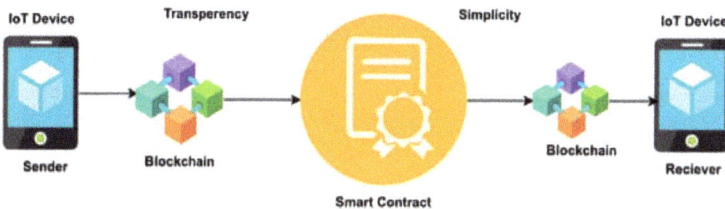

Figure 14.9   Smart contract [20].

• *Privacy*: This refers to the protection of data stored over the Blockchain from unauthorized access. Privacy is crucial for individuals who want to protect sensitive information, such as financial or medical data, from unauthorized access or manipulation. It ensures that only authorized parties can access and manipulate the data, which enhances privacy and prevents unauthorized access or tampering.

- *Trust*: Any industry in which transactions require a permanent record and trust from all parties can benefit from and potentially use Blockchain technology. Process-heavy practices, such as real estate title issuances and transfers, contract execution, insurance or trade finance, could similarly extend Blockchain technology to digitize and mechanize current processes. Blockchain is challenging the status quo of the central trust infrastructure currently prevalent in the Internet towards a design principle that is underscored by decentralization and transparency.

- *Interoperability*: This is the ability to share data and other information across multiple Blockchain systems as well as networks. This function makes it simple for the public to see and access the data across different Blockchain networks. This function also offers a range of diverse functionalities, for example, cross-chain transactions. Interoperability between Blockchains will lead to the development of added capabilities between Blockchains.

- *Automation*: It is possible to achieve a higher level of automation along the supply chain through Blockchain technology. Smart contracts can be verified and validated through the Blockchain system automatically.

14.8 CHALLENGES

While Blockchain offers astounding developments across sectors, it is still a work in progress. The main challenges faced by Blockchain technology include scalability, energy consumption, and regulatory issues. For all its promise, Blockchain has delivered surprisingly little. Its impact on the environment and sustainability have always been the topic of conversation. A lot of businesses still do not understand why they need Blockchain. Other challenges include the following [21]:

- *Complexity*: Blockchain seems complex and is still not widely or systematically understood. Many people do not understand exactly how Blockchain can help them, their company, or even society at large. The complexity of Blockchain technology, coupled with the prevalence of misinformation, has contributed to a general misunderstanding of how it works.

• *Regulation*: A hurdle Blockchain needs to overcome involves regulating the use of Blockchain technologies as well as related cryptocurrencies. This can give rise to new types of regulatory agencies that are going to manage the technology in different ways across the globe.

• *Demand for Skills*: Demand for cryptocurrency and Blockchain related expertise has increased exponentially in the last year. As the Blockchain technology grows, so does the need of skillful, highly trained individuals. With Blockchain technology still relatively new, there are limited numbers of Blockchain engineers and specialists. The demand for skills within the sector surpasses the current supply available. Universities and colleges that are able to accommodate new advances in decentralized technology will win the minds of the students who need to embrace these skills for the future.

14.9 CONCLUSION

Blockchain technology has revolutionized the way we store, secure, and transfer data across various industries. Without a doubt, Blockchain is one of the most valuable emerging technologies. While Blockchain offers astounding developments across sectors, it's still a work in progress. Blockchain is newly born technology, and it needs room for growth. Blockchain technology is poised to take over the way we work. The opportunity for people to deal freely will in fact generate opportunities that were unforeseeable before. Despite Blockchain being at the peak of its popularity, the job market has a lack of Blockchain engineers and specialists.

The Blockchain technology is at the height of its popularity even though it is still in the early stage. The technology is already impacting society and business on many levels. The world economy is getting ready for the Blockchain revolution. Many industry analysts have been charting possible paths for the future of Blockchain. Forecasts for the future of Blockchain focus on its increasing integration with emerging technology trends. Blockchain technology has made significant strides in its development and widespread adoption in recent years. There is no sign of it slowing down. It is here to stay. However, for Blockchain to continue to present new features and remain relevant, it is critical for Blockchain capabilities to evolve and integrate with newer technologies. More information about the future of Blockchain can be found in the books in [22-41] and the following related journal: Future Internet.

REFERENCES

[1] S. Banaeian, "Blockchain and its derived technologies shape the future generation of digital businesses: A focus on decentralized finance and the Metaverse," Data Science and Management, vol. 6, no.3,September 2023, pp. 183-197.

[2] M. N. O. Sadiku, U. C. Chukwu and J. O. Sadiku, "Future of Blockchain," International Journal of Human Computing Studies, vol. 5, no. 9, September 2023, pp.11-19.

[3] M. N. O. Sadiku, Y. Wang, S. Cui, and S. M. Musa, "A primer on Blockchain," International Journal of Advances in Scientific Research and Engineering, vol. 4, no. 2, February 2018, pp. 40-44.

[4] "Layered structure of the blockchain architecture,"

https://subscription.packtpub.com/book/data/9781789804164/1/ch01lv1l1sec07/layered-structure-of-the-blockchain-architecture

[5] M. Redka, "Future of the Blockchain technology: Use cases, risks and challenges," November 2021,

https://mlsdev.com/blog/the-future-of-the-Blockchain-technology-use-cases-geographical-expansion-potential-risks-and-challenges

[6] S. Depolo, "Why you should care about Blockchains: The non-financial uses of Blockchain technology," March 2016,

https://www.nesta.org.uk/blog/why-you-should-care-about-Blockchains-non-financial-uses-Blockchain-technology

[7] "The future of Blockchain: Applications and implications of distributed ledger technology," Unknown Source.

[8] M. Iansiti and K. R. Lakhani, "The truth about Blockchain," Harvard Business Review, Jan./Feb. 2017.

https://hbr.org/2017/01/the-truth-about-Blockchain

[9] W. T. Tsai et al., "A system view of financial Blockchains," Proceedings of IEEE Symposium on Service-Oriented System Engineering, 2016, pp. 450-457.

[10] L. Qiao et al., "Can Blockchain link the future?" Digital Communications and Networks, vol. 8, no. 5, October 2022, pp. 687-694.

[11] M. N. O. Sadiku, U. C. Chukwu, and J. O. Sadiku, "Future

of Internet of things," Horizon: Journal of Humanity and Artificial Intelligence, vol. 2, no. 5, 2023, pp. 632-640.

[12] "Artificial intelligence, Blockchain, and the future of the world," September 2022, https://medium.com/@Aiwork/artificial-intelligence-Blockchain-and-the-future-of-the-world-2fa5a121ee69

[13] "The future of AI and Blockchain technology & how it complements each other? https://www.turing.com/kb/how-Blockchain-and-ai-complement-each-other

[14] "The future of Blockchain technology: This is where it will be implemented first," https://www.zmescience.com/science/future-Blockchain-applications/#:~:text=naturally%2c%20finance%20is%20the%20industry,offering%20crypto%20as%20tradable%20assets.

[15] "Future of Blockchain: How will it revolutionize the world in 2022 & beyond!" November 2021, https://www.europeanbusinessreview.com/future-of-Blockchain-how-will-it-revolutionize-the-world-in-2022-beyond/

[16] "10 popular types of cryptocurrency and how they work," February 2023, https://n26.com/en-eu/blog/types-of-cryptocurrency

[17] "What is a non-fungible token? Beginner's guide to NFTs," May 2021, https://learn.bybit.com/crypto/what-is-a-non-fungible-token/

[18] "What is the future of Blockchain technology by 2025?" April 2021, https://lnct.ac.in/future-of-Blockchain-technology-by-2025/

[19] "10 Reasons why Blockchain technology is the future," https://www.geeksforgeeks.org/why-Blockchain-technology-is-the-future/

[20] T. Alam, "Blockchain-based internet of things: Review, current trends, applications, and future challenges," Computers, vol. 12, no.1, 2023.

[21] J. Godsil, "The future of jobs in Blockchain demands a flexible approach to education," October 2021, https://www.blockleaders.io/leaders/the-future-of-jobs-in-Blockchain-demands-a-flexible-approach-to-education

[22] P. Vigna and M. J. Casey. The Truth Machine: The Blockchain and The Future of Everything. Picador, 2019.

[23] M. K. Deep, The Future of Blockchain in Banking. GRIN Verlag, 2018.

[24] S. Mahankali, Blockchain: The Untold Story: From Birth of Internet to Future of Blockchain. BPB Publications, 2019.

[25] A. Kim, How Blockchain will Change the World : A Beginner's Guide To Blockchain Technology, Bitcoin, and the Future of Financial Technology. Independently Published, 2020.

[26] E. McFarland, Blockchain Wars: The Future of Big Tech Monopolies and the Blockchain. Evan McFarland, 2021.

[27] N. Mehta, A. Agashe, and P. Detroja, Blockchain Bubble or Revolution: The Future of Bitcoin, Blockchains, and Cryptocurrencies. Paravane Ventures, 2019.

[28] R. B. Seymour, The Blockchain Future: Bitcoin, Cryptocurrency, Blockchain Technology, De-centralised Ledgers, Smart Contracts, Crypto Wallets, NFTS and Web 3.0. What ... Do in the Real World Now and in The Future! Independently Published, 2021).

[29] P. Vigna and M. J. Casey, The Truth Machine: The Blockchain and the Future of Everything. Picadeor, 2019.

[30] A. Omar, The Future with Blockchain - Part 1. Amazon Digital Services LLC - KDP Print US , 2020.

[31] M. Train, Bitcoin: Predicting the Future of Blockchain Cryptos and Altcoins. Self Publisher, 2019.

[32] D. R. Okuto, The Future of Blockchain and Cryptocurrency. Fortugno, 2021.

[33] A. Agashe, P. Detroja, and N. Mehta, Blockchain Bubble Or Revolution: The Future of Bitcoin, Blockchains, and Cryptocurrencies. Paravane Ventures, 2019.

[34] M. Trainston, Bitcoin: Predicting the Future of Blockchain

Cryptos and Altcoins. Self Publisher, 2019.

[35] E. A. Ross, Cryptocurrency and Blockchain: Bitcoin Financial History and the Future of Blockchain Technology Blockchain Overview with Bitcoin Success Stories That Will Blow Your Mind. Eloise a Ross, 2022.

[36] E. Vandamme, The Future of Blockchain Technology. SSRN, 2019.

[37] J. C. Giancarlo, CryptoDad: The Fight for the Future of Money. Wiley, 2021.

[38] E. S. Prasad, The Future of Money: How the Digital Revolution Is Transforming Currencies and Finance. Harvard University Press, 2021.

[39] B. Lakeman, Blockchain Revolution: Understanding the Crypto Economy of the Future. A Non-Technical Guide to the Basics of Cryptocurrency Trading and Investing. SparkEnlite LLC, 2018.

[40] A. Wright, Blockchain: Uncovering Blockchain Technology, Cryptocurrencies, Bitcoin and the Future of Money (Blockchain and Cryptocurrency as the Future of Money, #1) Blockchain and Cryptocurrency Exposed. Alan Wright, 2017.

[41] I. Williams, Cross-Industry Use of Blockchain Technology and Opportunities for the Future. IGI Global, 2020.

INDEX

www.ingramcontent.com/pod-product-compliance
Lightning Source LLC
Chambersburg PA
CBHW040846210326
41597CB00029B/4750